Gender and Drone Warfare

This book investigates how drone warfare is deeply gendered and how this can be explored through the methodological framework of 'Haunting'.

Utilising original interview data from British Reaper drone crews, the book analyses the way killing by drones complicates traditional understandings of masculinity and femininity in warfare. As their role does not include physical risk, drone crews have been critiqued for failing to meet the masculine requirements necessary to be considered 'warriors' and have been derided for feminising war. However, this book argues that drone warfare, and the experiences of the crews, exceeds the traditional masculine/feminine binary and suggests a new approach to explore this issue. The framework of Haunting presented here draws on the insights of Jacques Derrida, Avery Gordon, and others to highlight four key themes – complex personhood, in/(hyper)visibility, disturbed temporality and power – as frames through which the intersection of gender and drone warfare can be examined. This book argues that Haunting provides a framework for both revealing and destabilising gendered binaries of use for feminist security studies and International Relations scholars, as well as shedding light on British drone warfare.

This book will be of interest to students of gender studies, sociology, war studies, and critical security studies.

Lindsay C. Clark is an ERC Research Fellow at the University of Southampton, UK.

Routledge Studies in Gender and Security

Series Editors: Laura Sjoberg
University of Florida
and
Caron E. Gentry
University of St. Andrews

This series looks to publish books at the intersection of gender studies, international relations, and Security Studies. It will publish a broad sampling of work in gender and security – from private military companies to world wars, from food insecurity to battlefield tactics, from large-n to deconstructive, and across different areas of the world. In addition to seeking a diverse sampling of substantive work in gender and security, the series seeks a diverse author pool – looking for cutting-edge junior scholars alongside more established authors, and authors from a wide variety of locations and across a spectrum of backgrounds.

Gender and the Genocide in Rwanda
Women as rescuers and perpetrators
Sara E. Brown

Gendering Military Sacrifice
A Feminist Comparative Analysis
Edited by Cecilia Åse and Maria Wendt

NATO, Gender and the Military
Women Organising from Within
Katharine A.M. Wright, Matthew Hurley and Jesus Gil Ruiz

Gender and Drone Warfare
A Hauntological Perspective
Lindsay C. Clark

For more information about this series, please visit: www.routledge.com/Routledge-Studies-in-Gender-and-Security/book-series/RSGS

Gender and Drone Warfare

A Hauntological Perspective

Lindsay C. Clark

Routledge
Taylor & Francis Group

LONDON AND NEW YORK

First published 2019 by Routledge

2 Park Square, Milton Park, Abingdon, Oxon, OX14 4RN
605 Third Avenue, New York, NY 10017

Routledge is an imprint of the Taylor & Francis Group, an informa business

First issued in paperback 2020

British Library Cataloguing-in-Publication Data
A catalogue record for this book is available from the British Library

Library of Congress Cataloging-in-Publication Data
A catalog record has been requested for this book

ISBN: 978-1-138-58027-5 (hbk)
ISBN: 978-0-367-78605-2 (pbk)

Typeset in Times New Roman
by Wearset Ltd, Boldon, Tyne and Wear

For David – for all the adventures

Contents

Acknowledgements

This book has been a labour of love and is woven out of frustration, interest, expletives, wine, debate, argument, and red pen marks. I owe my thanks to a myriad of wonderful humans who helped make it possible.

In the beginning ... there was Nicholas Wheeler and Nicki Smith, thank you for getting me thinking about the intersection of drone warfare and gender, and then for getting me through the PhD process that is the backdrop for this book! I have been fortunate to benefit from both of your excellent supervision. Thank you also to Penny Griffin and Laura Shepherd for generously sharing their time with me and for helping shape how I approached the project during my stint at UNSW Sydney. Thanks are due to the funders of my PhD, without whom the project that underpins this book could not have taken place: my thanks to the Economic and Social Research Council and the Royal Aeronautical Society (through the auspices of Peter Gray). To my fellow ICCS PhD colleagues, along with Jonna Nyman, Adam Quinn, and David Norman: Thank you for all the debates (and pizza and wine)!

The writing of this book from the original PhD project has been an involved process. I extend my thanks to Toni Erskine for providing me with both the environment in which I could complete this book and for such insightful feedback on every section she read. I would also like to thank the Humanities and Social Science department at UNSW Canberra for the post-doctoral fellowship under which I completed this. My thanks must go to Rhiannon Neilsen, who helped me re-think how I structure complex personhood and who discussed with me all the minutiae along the way (and all the yoga)! Thank you to the members of the International Ethics Research Group who so thoughtfully critiqued my theoretical chapters and helped me to improve.

I thank Laura Sjoberg and Caron Gentry for their interest in including this work in the Routledge Gender and Security Studies series and the anonymous reviewers for comments that were challenging and uplifting in equal measure.

Finally, I want to take a moment to thank all of the individuals who kindly agreed to be interviewed for this project. I cannot name you, but you know who you are. Thank you for taking the time to speak to me, for sharing your experiences with me. You have enriched my research experience and the overall book in ways I did not expect.

Acronyms

CIA	Central Intelligence Agency (USA)
DARPA	Defense Advanced Research Projects Agency (USA)
FOB	Forward Operating Base
GCS	Ground Control Station
GWOT	Global War on Terror
IED	Improvised Explosive Device(s)
ISIL/ISIS	Islamic State
ISR	Intelligence, Surveillance and Reconnaissance
ISTAR	Intelligence, Surveillance, Targeting and Reconnaissance
MALE	Medium Altitude, Long Endurance
MOD	Ministry of Defence (UK)
MP	Member of Parliament (UK)
MQ-1/MQ-9	Designation for Predator and Reaper respectively
NATO	North Atlantic Treaty Organization
NCOs	Non-Commissioned Officers
NGOs	Non-Governmental Organisations
PTSD	Post-Traumatic Stress Disorder
RAF	Royal Air Force
ROE	Rules of Engagement
RPA	Remotely Piloted Aircraft
RPAS	Remotely Piloted Aerial System
UAV	Unmanned Aerial Vehicle
UOR	Urgent Operational Requirement
USAF	United States Air Force

Introduction
(Dis)embodied warfare is ghostly

> He explained to me the odd reality of making his kid's lunch in the morning, kissing his wife goodbye, then driving to work in traffic, saying hello to his co-workers on base, and then passing through an aluminium door that miraculously transported him to the bloody battlefield of Iraq.
>
> (Rogoway, 2015)

> Every decision you made was either … somebody living, saving somebody or somebody dying. And you walk into your house and you're trying to figure out whether your daughter is going to wear a blue tutu or a pink tutu and the disconnection is astounding. It's just, it's … it's amazing.
>
> (Black, 2013)

These quotes reflect some of the strangeness of being part of a Reaper drone crew, in particular the constant cycling back and forth between the war front and the home hearth. They weave into other narratives of simultaneous geographical distance meshed against the curious intimacy created through extensive surveillance. The life of the 'Other' on the ground reflects the gaping cultural gulf between the crews and their 'targets' whilst at the same time the similarities of daily life echoes the crews' own lives in those marked out as 'the enemy'. The introduction of military technologies, like drones, is a topic that has produced a huge literature illustrating how these technologies affect war from a range of perspectives: ethical, legal, strategic, political, and human.[1] Technology in war has changed the way that war functions, changed the interactions between opposing sides, changed the way individuals are injured and the way that they die, changed how those deaths are interpreted, changed how those who inflict death are interpreted, and a myriad of other elements besides.[2] There are many, many, different kinds of military technologies that invite research and consideration. For example nuclear weapons continue to draw investigation, as do landmines and cluster munitions (Futter, 2011; Anderson, 2000; van Woudenberg, 2008). Interest has also been directed towards the potential of non-lethal weapons (Kaurin, 2010; Orbons, 2012), the potentially warscape-altering impact of cyberwarfare (NATO, 2003; Gartzke, 2013), and the increasing reliance on robots and robotics (Anderson and Waxman, 2012; Arkin, 2015; Doyle, 2013). It is this final category that has captured the public attention and imagination in recent years, exploding into a heated debate, and therefore it is military robots that this book will investigate, looking at a specific iteration: the armed drone. In investigating the implications of the use of armed drones, the book addresses one specific perspective which has, within traditional security studies perspectives on military technologies, been underappreciated, and that is the *gendered* implications of the use of armed drones.

Feminist scholars have, as will be outlined in greater detail in the following chapter, illuminated the ways in which war and gender are co-constituted, mapping the many diverse ways in which 'gendering' security studies is essential to enhancing our understanding of warfare in all its various iterations.[3] Similarly, important feminist interjections consider the work that gender does in the development of different technologies of war. Interpretation of the phallic symbolism of the thrust of the spear or sword has been expanded into explorations of the accusations of cowardice that accompanied the advent of the longbow through to investigations of the sexualised language used by US 'nuclear defence intellectuals', and even to the emancipatory potential of the cyborg (Ehrenreich, 1998; Lash, 1995; Cohn, 1987; Haraway, 1991). Therefore, feminist theorising has sketched, suggested, and argued ways in which we can begin to better understand what military technologies are doing *to* gender and *through* gender. Outside of feminist theorising there is a tendency to see technology as gender-neutral, and yet feminist scholars such as Carol Cohn (1987; Cohn and Ruddick, 2003), Mary Manjikian (2014), and Lauren Bayard de Volo (2013) have definitively shown the ways in which technology is gendered.

The quotes that started this chapter encapsulate some of the ways in which Reaper crew worlds, and the narratives about those worlds, are gendered. The breadwinner father, the stay-at-home wife, the daughter going to ballet; all reflect some of the most accepted discourses about gender appropriate behaviours. Add to this that the 'breadwinner father' is going to war, performing the 'most' masculine of roles and the stay-at-home wife who replicates Elshtain's (1995) beautiful soul is relegated to the private sphere, and the gendered discourses are rendered explicit. Despite this, commentators have criticised Reaper crews for eroding the masculinity of the 'warrior' as a result of the apparent 'risklessness' of their roles, whilst at the same time others have raised concerns that the drone represents the techno-fetishisation of the contemporary Western military, a cultural identity associated with hyper-masculinity (Vallor, 2013; Kunashakaran, 2016). How is it that these crews can be perceived as representing both sides of the gendered binary? And what is the relationship between technologies of war and gender? It is these questions that I explore in this book.

Existing interjections on this topic have explored the ways in which drone warfare can be constructed as a feminising influence on Western militaries, and understood as representing their hypermasculinisation as a result of techno-fetishisation.[4] In addition, drone warfare has been explored as a queer phenomenon in that it disrupts the heteronormative structures that organise traditional narratives of warfare.[5] This latter approach reveals the instabilities, contradictions, and inherent 'queerness' within those structures. Cara Daggett's perspective offers an important insight into the value of recognising the fluidity/instability of gendered binaries, noting that 'warrior archetypes that order the act of killing ... [are] defined against both the feminine and the queer' (Daggett, 2015, p. 362). These pieces constitute the background of debates against which this book is set.

Beginning with the arguments about feminisation, and (several miles) outside of feminist theorising, Martin Van Creveld's position on the connection between the 'feminine' qualities and the degradation of the Western military reflects his perspectives on the importance of keeping military spaces as bastions of 'pure' masculinity and manliness (Van Creveld, 2013b, 2013a). Whilst I fundamentally disagree with Van Creveld on these perspectives, his views provide a useful starting point from which to illuminate the gendered backdrop of the concerns of those who view armed drones as evidence of the decline of the militaries who use them. What Shannon Vallor (2013) describes as a concern about the 'moral deskilling' of the armed forces, can be viewed instead as concern about the erosion of military masculinity through the distancing of crews from the battlefield, and the attendant reduction in risk to life and limb.[6] The various mocking references to 'cubicle warriors' and 'Dilbert goes to war' suggest, rather than ethical concerns about the way the technology is used, a discomfort with the erosion of the fetishised heroic warrior figure – the ultimate construction of masculinity who risks his (strong) body to protect the 'womenandchildren' at home (*The Economist*, 2014; Mayer, 2009; Crane-Seeber, 2016; Enloe, 1989). Therefore some commentators have argued that discomfort with remote warfare instead 'rests in part upon a hierarchy of masculinity and its interactions with and implications for our understanding of "combatant" and "war hero"', and, I argue, 'warrior' (Bayard de Volo, 2013, p. 29). The question that emerges is: with the advent of distancing war technologies that apparently reduce the need for 'honour and valour' in war, how can we differentiate between undesirable killing and that which is worthy of our respect and value (Sauer and Schornig, 2012, p. 373)? And if the 'warrior' is no longer socially valuable what does this mean for a hegemonic form of masculinity?

Whilst the concerns about the feminisation of the militaries who use armed drones has tended to come from scholars outside of feminist theorising, those who identify themselves within those ranks have tended to argue that the use of armed drones has the potential to have the opposite effect; that is, an increasing hyper-masculinisation. Sumita Kunashakaran argues that whilst it is possible to see the skills that Reaper crews require as 'more feminine' than some other members of the military, the end result of the use of distancing technologies like drones is 'hyper-masculinity' (2016, p. 31). The physical distancing from the battlefield, Kunashakaran (2016) claims encourages emotional distancing, a trait she equates with masculinity. Similarly, both she and Caroline Holmqvist (2013) equate the physical 'body' of the drone with the hyper-masculine – representing the hard, erect, impenetrable body of the ideal soldier; which provides the means through which drone crews perform their military masculinity. Lauren Bayard de Volo looks beyond the crews to argue that the use of drones in the 'Global War on Terror' (GWOT, now more euphemistically referred to as 'Overseas Contingency Operations' (Burkeman, 2009)) represents 'a paternalistic expression of rescuing a feminized region', with the 'penetration' of sovereign space representing the 'demasculinisation' of areas in which drones are used (2013, pp. 12, 13; see also Shepherd, 2006). Feminist scholars have also pointed to the increasing importance of 'technical proficiency' as indicators of a 'different kind of male activity', which is then expanded onto the world stage and establishes the state with technology as empowered over the one without, which is then feminised and can be bullied, cowed and intimidated by technology (Manjikian, 2014, p. 53; see also Kontour, 2012; McKinley *et al.*, 2011). Similar arguments about gendered discourses surrounding the use of technology are also explored by Cristina Masters, who notes that pre-existing 'military discourses have constructed the cyborg soldier', who in this instance are the aircrew tasked with operating the Reaper (2005, p. 113). These discourses are not easily destabilised and therefore, Masters (2005) and Manjikian (2014) argue, are likely to embrace novel war technologies in ways that reinforce existing gendered binaries.

Whilst these authors make often convincing arguments, in many of these apparently opposing accounts there are acknowledgements of elements that do not fit neatly into the narratives that the authors are trying to create. Interactions with military technologies and the gendered implications of these interactions are rarely simple or one sided. For example, whilst arguing that the end result is hyper-masculinisation, Kunashakaran notes that 'drones do indeed open up new avenues for more "feminized" skills' and that the traditional image of the physically strong war hero is 'also challenged among UAV [unmanned aerial vehicles] operators' (2016, pp. 32, 35). The emotional remoteness (coded masculine) of the 'drone stare' Holmqvist argues 'furthers the subjugation of those marked as Other' but then goes on to note that this narrative is disrupted by the technical capacity to see in detail which creates an opposing 'sense of proximity' provoking an emotional response in drone crews (2013, p. 452). Similarly, whilst the drone operates as a 'paternal' figure in International Relations the crews cannot demonstrate their masculinity in the traditional fashion of 'mastering fear or resisting the impulse to flee' from danger (Bayard de Volo, 2013, p. 11). Queer theorising by scholars like Cara Daggett and Lauren Wilcox has enabled a different way of thinking about gender and drone warfare, such that it disrupts the heteronormative structures that organise traditional narratives of warfare. As such, Daggett (2015) argues that drone warfare is uniquely disorientating. However, whilst Millar and Tidy argue that Daggett's analysis 'can be read as concerning the maintenance of the heroic soldier myth and the production of martial violence as "combat"' (2017, pp. 149, 154), there is space for a stronger (but also queer-inspired) engagement with the concept of the warrior itself and how this is historically/socially/culturally situated. Therefore, this book is interested in a different way of understanding the 'heroic soldier myth', and an attempt to find a framework to understand *how* military technologies can simultaneously destabilise and (re)inscribe gender discourses.

Ghosts

To tell the story that is at the core of this book it is necessary to find a framework and methodology that can accommodate *both* sides of the gender binary. At the same time, it requires a framework and methodology that embraces complexity rather than arguing that the use of military technologies, like armed drones, result in *either* feminisation *or* hyper-masculinity. In this book I use the framework of 'Haunting'[7] and the methodology of ghost-hunting. I understand Haunting as a vocabulary to 'communicate the depth, density, and intricacies of the dialectic subjection and subjectivity' and the ghosts as 'conceptual metaphors' or 'deconstructive logic[s]' that do the work of building in the complexity of day-to-day life (Gordon, 2008, p. 8; Blanco and Peeren, 2013, p. 1; Bal, 2010, p. 10). Outlined in greater detail in Chapter 2, as part of the wider 'spectral (re)turn', Haunting does not focus on the occult or parapsychology, but rather operates as a means of 'escape[ing] the totalizing logic' which is often applied to issues of gender and sexuality (Peeren, 2014, p. 10). Escaping totalising logic means paying attention to the things that *don't* fit in the ways we expect them to, engaging with non-linearity, with things that trouble us, with hunches, intuitions and other 'deviant knowledges' (Weber, 2014, p. 598). In this book, I use Haunting to foreground complexity and nuance, highlighting four core components of the framework: (1) complex personhood, (2) in/(hyper)visibility, (3) disturbed temporality, and (4) power. Through the prism of Haunting, I explore these issues and build my analysis around the interaction between gender and military technology in order to engage with, and trouble, binaries such as silence *and* scream, absence *and* presence, visibility *and* invisibility, masculinity *and* femininity; and do so in a way that enables the opposing sides of the binary to co-exist (Blanco and Peeren, 2013, p. 1). The core contribution of this book is in the application of the framework of Haunting to these topics, through which I argue, it is possible to gain more nuanced and detailed understandings of the way that history, social structures and technologies interact. Doing so is ethically important, because, even though it will not result in 'a more tidy world' it could help us to create 'one that might be less damaging' (Gordon, 2008, p. 19).

From the framework of Haunting comes the methodology of the ghost hunt. This aims to draw attention to (cob)webs of power that affect subjects, understanding those individuals as complex and as such enmeshed in a complex social, psychological, cultural, and historical context that can dis/empower. Feminist scholars have repeatedly illuminated the way in which gender acts in dis/empowering ways, and drawn attention to the complexity of the work that gender *does* and the multiple sites in which it functions (for example Enloe, 1983; Peterson, 1992; Shepherd, 2008; Steans, 2008; Guerrina and Wright, 2016). I outline in Chapter 2 the ways in which Haunting and the ghosts provide a useful means of exploring how gender functions. Feminist work on military technologies has utilised novel methodologies and figurations such as the cyborg and the monster (Haraway, 1991; Rayner, 1994; Creed, 1993; Ussher, 2006; Balsamo, 1996). However, in relation to technologies that are so intimately associated with death and killing, it feels particularly apt to utilise Haunting and ghost hunting (Auchter, 2014; Gordon, 2008). What could be more appropriate for a technology called Reaper? Additionally, gender has often been (necessarily and usefully) tied up with discussions of the physical body, but through the figure of the ghost it is possible to engage with a discussion of gender that can connect the physical body with gendered discourses and binary logics (Peterson and Runyan, 2010).

Ghosts and queer logics

The final piece of the framework of Haunting is the addition of what Cynthia Weber (2014, 2016) has called 'queer logic(s)'. In order to explain queer logic(s), it is necessary for me to take a brief detour to explain the importance of binaries to thinking about gender (and therefore also military technologies). Feminist scholars such as V. Spike Peterson and Ann Sisson Runyan (2010) have drawn attention to ways that masculinity and femininity are constructed in opposition to one another, so that what is masculine is what is not feminine and vice versa. Certain personal, political, and structural attributes or performances are constructed as being either masculine *or* feminine; and as feminist scholars have shown, this has myriad implications for issues of importance to security studies.[8] In addition to being constructed in opposition to one another, the masculine/feminine dichotomy is hierarchical, with the feminine devalued in preference for the masculine. In relation to military masculinities what is constructed as desirable behaviour/identity are things like (for example) physical strength and emotional reserve (coded as masculine), whereas physical weakness and emotionality is construed as feminine and undesirable.[9] The power of these kinds of binaries has been investigated by feminist scholars, who have demonstrated their implications in conflict situations.[10] Challenging these binaries has become part of the important project of feminism, and indeed an important part of this book. I utilise queer logic(s) that challenge the oppositional nature of the masculine/feminine binary (amongst others), illustrating the extent to which these binaries are constructed and mutable. As I flesh out in Chapter 2, applying a 'queer logic(s)' disrupts the either/or structure of binaries, opening up the possibilities, destabilising the idea of a *or* b to include a *and/or* b (Weber, 2014, 2016). As such, the use of queer logics draws attention to instabilities and contradictions at work in gendered binaries in relation to war and war technologies because the experiences of the crews are always, already 'exceeding all binary opposites' (Harris, 2015, p. 17).

Understanding ghosts as embodying queer logics and applying this logic to the use of military technologies and gender reveals a ghost that helps to illuminate the tangle of influences on military masculinity (and its relationship with femininity), and this is the ghost of the warrior. As a conceptual metaphor, or 'figuration' (Haraway, 1997), the ghost of the warrior illuminates the varied and vibrant parentage of military masculinity and its inherent instabilities and contradictions which suggest a different possible relationship with femininity than simple opposition. In Chapter 3 I explore how this coherent figure can and does contain contradictions that in no way impair our ability to understand what the figure *is* and what it *means*. As such, the ghost of the warrior is a performative 'trope' rather than a literal representation of the warrior (Weber, 2016; Haraway, 1997). A literal representation would be impossible given the way the ghost comprises a multitude of translucent layers of meaning, which sometimes appear to be mutually exclusive. Therefore, my deployment of the ghost of the warrior is 'the act of employing semiotic tropes that combine knowledges, practices, and power to shape how we map our worlds and understand actual things in those worlds' (Weber, 2016, Kindle location 663).

The ghost of the warrior is the discursive backdrop against which I examine the gendered implications of the use of military technologies, laying out in Chapter 3 the 'traditional' (which is not to imply 'true') context in which these technologies are operated. Utilising the ghost of the warrior in this manner enables me to draw out the extent to which being a warrior comprises so much more that simply *being* a warrior. Much in the way that Holmqvist utilises Elaine Scarry's argument that in war 'fighting always exceeds "fighting"' (Holmqvist, 2013, p. 537), so too does the warrior exceed the 'warrior'. War is centrally constituted by what it means to be a warrior, and that war cannot be understood without recourse to who or what the warrior means. The interaction between the warrior and war technologies is important beyond the singular strategic or legal implications, but requires an understanding of the human experience of war, and this always means including the gendered implications.[11] Similarly, the ghost of the warrior represents the *excess*, that is '[t]he specter stands for that which never simply is' (Peeren, 2014, p. 10) but rather represents the way that queer logic(s) function to destabilise what is apparently masculine (and/)or feminine. This book is concerned by the way that these logics play out in the introduction of novel technologies of war and centrally I argue that these can only be adequately understood by employing a framework and methodology that can encapsulate the complexity of the interaction(s) – in particular through the framework of Haunting. Understanding how these technologies are implicated in destabilising and (re)inscribing gendered discourses means engaging, and exploring how these are manifested in the specific, local, contexts in which each technology is used. To this end, I focus on the specific context of the use of armed Reaper drones by British crews in the Royal Air Force (RAF) in order to engage with the theoretical statement that 'life is complicated'[12] as it relates to the development of military technologies and the gendered narratives that (re)produce and challenge 'traditional' understandings of what it means to be a warrior (Gordon, 2008, p. 3).

Researching military technologies

Building a book on war technologies and gender from the threads of ghosts and queer logics reflects my stance as a poststructural feminist. Identifying myself as such means that I am concerned about the impact of gender on the lives of individuals and the consequences of structures of inequality that are based on arguments relating to sex and/or gender. The poststructural part of the statement means that I am concerned about the way that sex and gender inequalities play into intersectional inequalities – those based around race, class, ethnicity, sexuality and (dis)ability (inter alia Peterson and Runyan, 2010). The poststructuralism also means that I am concerned by the way that inequality (re)constitutes itself through narratives and discourse, stories that we tell ourselves about how people do, and should, act. Adopting a poststructural feminist stance has made me critical of claims of 'every/all', has made me interested in how the human stories at the 'bottom' or the margins of the international system speak to overarching structures of gender.[13] It requires that I try to think about 'situatedness' in creative ways in order to better reflect the intricacies of the environment in which we function and the ways that individuals understand themselves.

Given the limitations of language, there are few ways I can avoid creating boundaries when I speak of the problems of gendered binaries, and indeed I have needed to use parts of this book to sketch out what I have termed 'traditional' constructions of masculinity and femininity. However, following on from the early feedback and thinking reflexively about the work the words I put on the page are doing, I have tried to be explicit about where the construction of 'masculinity' or 'femininity' is problematic or counter to the aims of this project. As I note in the introduction to Chapter 3 on the warrior, these 'traditional' narratives are no more 'real' than non-traditional narratives, and the 'real' of history provides a plethora of examples that act as counter-claims. However, the 'traditional' gendered discourses of the warrior have a certain power, a power that disciplines and (re)produces certain violences that do not go away if we ignore them. Therefore, in keeping with the requirements of Haunting that we 'shine a light into the shadows', I have tried to sketch out the ghostly shapes of these 'traditional' narratives as a point from which to begin to contest their claims, and through that their disciplinary power.

Counter to the claims that researchers can be 'objective' and produce 'neutral' observations to form scientific answers to hypotheses, I view myself as having a role in the questions I asked, the data that I collected and the way that I interpreted it. These ethical commitments are in keeping Haunting which requires that I consider what I am *not* saying and what I am *not* hearing as much as those things that I am. It is a framework that is dedicated to trying to 'understand and write evocatively about some of the ways that modern forms of dispossession, exploitation and repression concretely impact the lives of the people most affected by them' (Gordon, 2011, p. 1). As explored in Chapter 2, such an ethical commitment cannot be met by claims to objectivity, but rather by an engagement with the sensuous, the emotional, and the strange. Some elements of this 'strangeness', in the case I consider, emerge through the narratives that are deployed around how to be, and what it means to be, part of the British armed forces, and how these narratives interact with developments in military technology.

It would not be excessive to argue that the introduction of armed drones has been one of the most contentious and debated developments in military techno-logy in recent years.[14] Drones represent a new iteration in airpower technology that situates the crews outside the aircraft and often at extreme distances from the area in which the aircraft is being flown. All of this adds up to a novel experi-ence for the individuals tasked with the drone's operation. The discussion of Reaper crews is particularly important because Reapers (and its predecessor Predator) are amongst the few armed drones in operation and are utilised by British and American crews.[15] Other armed drones have been used by the Israeli military and armed drones are in the process of being acquired or developed by many other states. The Reaper crews are therefore not just *flying* the drone from an extreme distance, they are also being tasked with deploying lethal force from the same distance. The crews are made up of three individuals: the pilot, the sensor operator, and the intelligence analyst, and together these individuals com-prise the focus of this study. We traditionally conceive of members of the military as demonstrating their status as warriors by risking life and limb in close combat, where the situation is 'kill or be killed'. As Reaper crews are at extreme distance from the 'theatre of war' they are experiencing war in a way that chal-lenges traditional masculine conceptions of what it means to be a warrior. These challenges to warrior identity are important because they are having an impact on the well-being of the crews and their families, as well as affecting recruitment and retention of crews.[16] The British case is of particular interest because at this time there is considerably less information available about these crews than American counterparts. The experiences of British crews is understudied and, in popular reports, lacking in nuance.[17] Given the recent lifting of export limitations on Reaper technology, other countries (such as Italy, Canada, and Australia) have begun looking to include these aircraft in their military inventories.[18] The British experience of using Reaper (and the experience of training to use Reaper with US crews) can therefore be instructive for policy makers in countries con-sidering making such purchases.

In order to better understand the implications of the use of armed drones on gendered discourses of war I employ a range of different kinds of data, and different 'discursive moments'. Much as Gordon utilises fiction as a means of investigating issues such racism, sexism, state violence, etc. (as discussed in more detail in Chapter 2), I engage with factual and fictional accounts: interview 'data', transcripts of debates in the British House of Commons, memoirs, car-toons, newspaper accounts and films. However, my starting point is interview data because there has been so much tendency within the press (both British and international) to rely on hearsay or to make unfounded assumptions about the lives of Reaper crews. Given the lack of access to Reaper crews there has emerged a strange tautological circuitry in both academic and journalistic articles: academic articles cite journalistic accounts which cite articles or other journalistic accounts. The pieces that pose as 'factual' become embroidered through the process of serial and tangential reconstruction and retelling. There is a lack of good quality open source data and therefore Reaper crews remain shrouded in secrecy. Even respected journalists who have tried to dig out 'real' stories have fallen foul of misinformation as Joseph O. Chapa's critique of a story by Mark Bowden illustrates:

> The story's protagonist, a '19-year-old American soldier' who entered Air Force basic military training straight out of high school, became an MQ-1 Predator crew member upon graduation. Reportedly, on his very first mission at the controls, the 'young pilot' observed a troops-in-contact situation on the ground. The 'colonel, watching over his shoulder, said, "They're pinned down pretty good. They're gonna be screwed if you don't do something."' The narrative goes on to describe the Hellfire missile strike and the psychological effect it had on the Airman. To a sophisticated military audience, the factual inconsistencies in this account are apparent. Air Force RPAs are crewed by Airmen, not Soldiers. The 19-year-old Airman (an enlisted rank) cannot be an Air Force pilot (an officer rating). The article claims that during his first time at the controls, this Airman finds himself on a combat mission in theatre. In reality, he would have become familiar with the controls at initial qualification training, prior to arriving at his first combat squadron. Furthermore, when colonels speak to Airmen about life-and-death combat decisions, they tend to do so in terms of direct orders rather than leading suggestions. How can Mark Bowden, notable historian and author of such well-received books as *Black Hawk Down*, commit such factual errors? The answer is simple. Information about Air Force RPA operations is rarely available—and when it is, it usually proves unreliable.
>
> (Chapa, 2014, pp. 29–30)

All of this is not to claim that the data that I use, or my use of it, is somehow 'better' or more objective than other accounts. Rather, I wanted to get as close to the people whose experiences I was interested in as I could. The interviews were between one and three hours and were conducted in person and via Skype. I provided my interviewees with some basic, open-ended questions to start with but largely let the interviewees drive the direction of the discussion, emphasising what *they* thought was important about their experiences of crewing Reapers. The interviews represent part of my means of engaging with an often-unheard part of the drone-discourses: the crews themselves. They provide some insight into the way that the crews understand their own roles, their position within the British military and their engagement with public debates. The people I interviewed did not, and do not, exist in a vacuum, but rather are woven, tangled and merged with all the background (and foreground) chatter about their roles. They are also complicated by the fact that they are more than their roles. They are confusing, messy, social, cultural beings. They live and breathe in contradictory, convoluted ways that are sometimes hard to make sense of.

Whilst critical military scholars[19] have sketched out the dangers of legitimising military violences by engaging directly with militaries, I follow Harriet Gray who argues that, although this is certainly a concern 'there also exist significant ethical and practical drawbacks to research which *does not* connect with the institution' (2016, p. 74). The risk is that without engaging it is easy to construct the military as monolithic, to neglect the fact that

> Institutions are made up of *people*, and thus, failing to pay critical attention to the complexities of the experiences and views of the people who constitute the military diminishes the potential power of our engagements.
>
> (Gray, 2016, p. 75)

Engaging with interviewees I was aware of a sometimes-conflicting need to do justice to the individuals who had so kindly given up their time to discuss their experiences with me and my desire to remain critical (although not necessarily negative) of the way that these crews lived their lives and the operations that they are tasked with. Reflecting on my experiences, I felt some kinship with Victoria Basham, who noted of her own fieldwork with the British military

> It was my first foray into 'the field' and I was worried I'd get it all wrong … I also had reservations about whether I should be conducting fieldwork with soldiers at all. My aim was to better understand how soldiers experience, and possibly context, social knowledge of gender [and] sexuality…. However, I was also well aware of the reluctance of some fellow scholars to engage with war and militaries for fear of legitimising militaristic practices.
>
> (2013, p. 2)

I, too, was worried about getting the field work 'wrong'. I was worried that through this project I would either fail to be ethical and fair in my treatment of the interview data or I would accidentally slide into becoming a drone apologist. However, by engaging with feminist scholarship and the idea of queer logics I have tried to present an account of the lives and identities of British Reaper crews that confirms their humanity and reveals some of the complexity of their experiences. They are not only Reaper crews: this is the aspect of life experience that they share, and it provides the backbone and motivation for the study, but this is, importantly, not all they are. They are also men and women, they are raced, and classed, they are wives/husbands/partners/single, they are parents and dog owners, they are car fanatics, lovers of baseball and/or football, and they are gamblers, drinkers, teetotallers, and they appreciate a good film on occasion.

The 'human-ness' of war, and of the individuals who undertake drone warfare, was brought home to me through the experience of conducting interviews. Through engaging with interviewees, I was forced to reflect on how they expressed their experiences, how I responded to those expressions, and how I was changed by the experience of doing so. My own humanness and my part in the political construction of the state that requires them to undertake war was all part of the questions that the experience of fieldwork raised. Visiting RAF Waddington emphasised more clearly than any written account could the closeness and the interaction between warfare and daily, domestic life. The incongruity, to me, of the children's playpark within the confines of the base served as an important reminder of the multi-faceted nature of the individuals I was engaging with. I was prompted to consider that their experiences of crewing Reapers marked out only a small part of their life, even at the same time as that small part made them 'research relevant' to me. Similarly, whilst searching for relevant participants I noticed how time spent as Reaper crew members was a remarkably short part of their work history. What I have tried to write, then, is something that draws attention to the fact that that coherence is only partial, that the 'sensuous'[20] knowledge that I drew from the shrugs of shoulders, the rolling of eyes, and a myriad besides is amorphous, spectral, and sometimes a tiny bit bizarre (and wonderful).

Reaper

Recognising the complexity of experiences and people is at the heart of the choice of the British case as the site of analysis. Whilst the US is undeniably the 'greatest'[21] user of armed drones, the British case represents an important and currently understudied case of the use of Reapers. As previously noted, there exists little in-depth academic research into the British use of Reaper drones, and although the newspapers include many different reference to the US use and to the experiences of US crews, there is much less time and space expended on how armed drones, specifically Reaper drones, are used by the British.

It is necessary here to explain my use of the term Reaper and to delineate what I mean when I speak about drones. The drones I am referring to in this book are remotely piloted aircraft (RPA). Whilst the term may be used for remotely operated ground and marine vehicles, it is most frequently used to describe flying vehicles where there is no pilot physically situated in the aircraft. There are a wide range of different types of drones. Drones may be the size of mosquitos or the size of jet liners, additionally they may be powered by propeller or have static wings.[22] Although it is true that the majority of drones are unarmed, it is without doubt that those which have the capacity to launch missiles strikes are the most contentious and have provoked the greatest amount of public debate and unease. In the UK the only armed variety of drone is the General Atomics Reaper (this was due to be replaced by a new variant, 'Protector' in 2020, but the start date has now been pushed to 2024).[23] Reaper is a medium altitude long endurance (MALE) drone, which was initially designed for reconnaissance missions, but has since been reconfigured to be able to carry munitions, specifically either Hellfire missiles or GBU 12,500 lb 'laser guided bombs' (Royal Air Force, 2016). Even though the majority of drones are small enough to hand-launch, with a wingspan of 20.12 m '12.5 feet of height and 36 feet of overall length', Reaper is much larger and requires individuals with specialist training to get off the ground (Royal Air Force, 2016).

When I use the term 'Reaper' in this book I do so in an attempt to refer with accuracy to the specific kind of drone I am speaking of. The term 'drone' is highly contested and, whilst there is certainly not space to go into the various revolutions of that debate, I would be foolhardy not to be clear that within British military circles the term is vehemently opposed.[24] The term drone is most often utilised by those who are critical of their use, rather than those who are actually tasked with using them. This project aims to shed light on the lives of those individuals whose jobs it is to use drones, and to convey this to a wider audience. I therefore have a dual responsibility to be fair to the individuals whose lives I am trying to illuminate and who kindly shared their experiences with me in interview, and the wider academic and public audiences who deserve to have the debate couched in terms that are clear, understandable, and appropriate. As such, I use the term 'Reaper' when speaking specifically about what the crews were piloting because this highlights the specificity of their experience and acts as a reminder that this is only one of many different types of 'drone'. I will use the word 'drone' when discussing the phenomenon generally because of its cultural and social importance and because this book aims to speak to an audience broader than the British or US militaries.

In order to engage more widely in the debate this book builds on the excellent work done by feminist scholars on the role of gender in the British military, understanding the institution as a site that shapes, moulds, changes, and reinforces dominant conceptions of military masculinity.[25] My interest in the lives and identities of British Reaper crews is not just about adding to the feminist security studies literature or security studies more generally, but goes beyond academia. Initial research has indicated that American Reaper crews are experiencing higher than expected levels of stress, burn out, and a novel kind of Post-Traumatic Stress Disorder (PTSD).[26] There is also some evidence that at least part of the distress these individuals experience relates to confusion over how their roles fit into the wider military, how they are perceived by their peers, and how they relate to their families.[27] Given that British Reaper crews are experiencing some of the same stresses as their American counterparts, as well as some unique to them, and given that the British government has recently pledged to increase the number of Reapers that the RAF operates, it is important to explore how the experiences of crewing this technology is affecting individuals and squadrons. This study does not aim to provide a way of making the situation 'better', but rather to outline how the experiences of British Reaper crews fit into the broader narratives of British military masculinity, with implications for policy development, recruitment, and retention.

This book focuses on the British use of Reaper; however, it is essential to sketch out a brief outline of American usage because the US is undoubtedly the most prolific user of the technology, and it is against the background of this usage that the British debate is situated. The intermingling of the British debate with US use of armed drones is wide-ranging and entrenched. For example, many critics of the use of drones have centred their arguments on the Bush and Obama Administrations' use of the technology in targeted killing campaigns.[28] As part of these campaigns drones armed with hellfire missiles have been deployed to strike (kill) specific individuals on the basis of either specific intelligence that they are an enemy of the United States (e.g. a member of Al Qaeda), referred to as personality strikes, or on the basis of patterns of life observation that indicate that they are likely to be engaged in anti-American activities (signature strikes).[29] These strikes have been undertaken by the US Air Force (USAF) but also by Central Intelligence Agency (CIA) operatives, fuelling debates about the appropriateness of civilians deploying lethal force and further complicating the already muddy waters surrounding who is and is not a combatant.[30]

Another criticism of US drone strikes is that they occur in countries, most notably in Pakistan, where the US is not officially 'at war'. In undertaking strikes in Pakistan, Afghanistan, Yemen, and Somalia, and targeting non-state actors the United States has expanded the boundaries of where acts of war (i.e. drone strikes) can take place, as '[p]lainly the United States is not at war with Pakistan' (Gregory, 2011, p. 241). Drone strikes now form a significant part of what was 'The Global War on Terror' (GWOT). However, whilst the Bush administration began the use of drone strikes as part of its efforts in the GWOT, it was the Obama administration that significantly increased the number and range of strikes (Serle, 2014). Despite pledging to end the 'war on terror' when he first entered office, '[d]uring his first term President Obama launched more than six times as many drone strikes as President Bush did throughout his eight years in office' (Boyle, 2013, p. 2; see also Dworkin and European Council on Foreign Relations, 2013).

Because of these strikes thousands of individuals have been killed (Serle, 2014). The US government has claimed that the majority of these individuals were high-profile terrorists, whose deaths were necessary in order to protect the American homeland from attack.[31] But critics argue that many of those killed were neither legitimate targets nor of strategic importance.[32] There has also been significant variance in reports on civilian casualties, complicated by reporters' awareness of the sensitivity of international and national audiences to the killing of innocents. For example, the Bureau of Investigative Journalism estimates civilian casualties of drone strikes in Pakistan between 424 and 969, in Yemen 174–225, in Somalia 10–58, and in Afghanistan 150–367 (TBIJ Drone Wars Casualty Estimates 19 December 2018).

Whilst the debate rages over whether or not drone strikes kill more or fewer civilians than terrorists, whether drone strikes create more terrorists than they kill, and whether or not the capacity to undertake strikes without the risk of losing pilots lowers the barriers to the use of force, there is no indication that the US has any intention of stopping or cutting down on the number of drone strikes.[33] It is this cacophony of strikes, debate, deaths and misinformation that (mis?)informs the British public's perspective of the use of Reaper drones. This is despite the fact that there are some significant differences between the way that the US uses the technology and the way it is utilised by the British military.

The British case

The British position is significantly different.

(Lee, 2015, p. 121)

The UK began with six Reaper aircraft (one of which was lost), then added an additional five in 2012. Since then up to 16 additional airframes have been ordered with the possibility of ten more.[34]

By 2012 '[t]he UK's current fleet of five Reaper drones has now flown nearly 54,000 hours in Afghan operations and fired their weapons 459 times' (Ross and Woods, 2012). British drones were initially used for intelligence gathering, reconnaissance, and force protection in Afghanistan. Since then their role has been expanded to include reconnaissance missions in Libya, Syria, and Iraq; and strike campaigns in Syria and Iraq where 2,767 missions have been flown by Reaper crews with 896 weapons released.[35] However, it is important to note that Reaper drones are not the only aerial vehicles being used by the British in these theatres; manned Tornado and Typhoon jets have also been deployed and Reaper strikes amount to 398 out of a total of 1,681 (2014–2017).[36] This information re-emphasises the way in which drones are used by the British military, indicating their position as one possible tool amongst many others. The majority of British Reaper usage has not taken the format of targeted killing campaigns, but rather fits into the more traditional understanding of Reaper as simply another part of the airpower toolkit.[37] British Reapers are flown under the same rules of engagement as manned aircraft and the same legal and ethical principles apply to any kinetic engagement (that is the deploying of any missiles). A redacted statement from the Squadron Leader of one of the two British Reaper squadrons notes that there are only three scenarios in which a Reaper crew may be authorised to use kinetic force. These are:

a   There is true belief that there is an imminent threat to life, e.g. Friendly Forces are in contact with enemy forces.
b   Real evidence of hostile intent to Coalition Forces e.g. enemy forces with weapons moving to firing points in close proximity to friendly forces.
c   Witnessing of a hostile act such as the active laying of an IED. (In the Matter of an Appeal to the First Tier Tribunal)

Despite this apparently limited use of force, issues of ethics and legality have become increasingly important since UK Reapers have been deployed outside of the traditional theatre of war to Libya, and Syria, particularly given the backdrop of US strikes.[38]

Given the contentious nature of US actions, the British government has sought to emphasise the difference between 'their' and 'our' use of drones: 'We consider that it is of vital importance that a clear distinction be drawn between the actions of UK Armed Forces operating remotely piloted air systems in Afghanistan and those of other States elsewhere' (House of Commons Defence Committee, 2014b, p. 13). Similar conclusions were reached by the members of the Birmingham Policy Commission report *The Security Impact of Drones*, which noted as one of its key recommendations: 'the case for making explicit UK policy on the legal and ethical considerations relating to RPA, including attitudes to US policy ...' (Birmingham Policy Commission, 2014, p. 82). Notwithstanding the desire to 'normalise' the use of British Reapers there is still a reluctance within the military to be more open in discussions about drones, as Chris Cole who runs the non-governmental organisation (NGO) Drone Wars noted:

> The real difference between the two types of aircraft becomes apparent when considering the reporting. While the UK is happy to report the number of Tornado aircraft taking part in strikes in Iraq and that they are flying from RAF Akrotiri in Cyprus, by contrast the UK refuses to give the number and location of UK Reaper drones taking part in strikes. This secrecy is echoed in reporting of the strikes.
>
> (Cole, 2015a)

In addition, articulating the difference between the British and American approaches is complicated by the sharing of intelligence and cooperation in various missions between the United Kingdom and the United States.[39]

There are indications that whilst the use of Reaper might be becoming normalised within the military, the British public remains concerned and somewhat sceptical. Indeed, the Defence Committee noted '[although] RPAS had made a significant contribution to operations in Afghanistan and Iraq, providing enhanced intelligence, surveillance and reconnaissance support in addition to weapons use' [there is] 'a sense of public disquiet' which needs to be addressed (Brooke-Holland, 2015; House of Commons Defence Committee, 2014a, p. 48). As this introduction and book goes on to explore, this 'disquiet' has implications for the crews who are tasked with operating Reaper drones.

Who/where are the crews?

Popular media accounts of Reaper crews have provided a variety of perspectives. Reaper crews, through their connection with technology and their distance from the theatre of war, confuse our traditional stories of what it means to 'go to war', a confusion that is reflected in the defensive, and sometimes derogatory terms and tones that are used in articles about the crews such as 'cubicle warriors', 'armchair killers', 'long distance warriors', 'bureaucratised killers', 'cyber-warriors', 'button pushers', 'warrior geeks', 'remote control warriors', 'remote soldiers', 'office warriors', and 'techno-warriors' which draw attention to some of the reasons for public discomfort over the use of armed drones (Asaro, 2013; Coker, 2007; Freeman, 2015; Alexander, 2015; Donnelly, 2005; Mulrine, 2012; *The Economist*, 2014; Calhoun, 2011; Royakkers and van Est, 2010). This kind of labelling matters because it constructs Reaper crews as something different from the usual air crewman/woman, as something outside of our normal narratives of the heroic/brave/sexy 'Top Gun' pilot.

Perhaps as a result of the negative perceptions of their roles, and the secrecy that the Ministry of Defence (MOD) has imposed regarding their operations, I was concerned that it would be difficult to engage in discussion with former crews. However, this was not the case, and many individuals were delighted to have an opportunity to 'set the record straight'. As with all research into the experience of ex-military individuals, I felt that building a rapport with my interviewees was particularly important because of the potentially distressing nature of some of the topics of discussion (including the impact of lethal strikes and psychological trauma). Therefore, by building up trust in the discussion and letting the interviewee lead the direction I tried to avoid forcing anyone to discuss topics that they were not comfortable with. The former crew members who I interviewed were in some cases still bound by the Official Secrets Act and therefore some of the examples that they gave me to illustrate their claims either had to be completely off record or presented in a vague or partly fictionalised manner. Where I have utilised this material, I have tried to be faithful to what I believe was the message behind what was told to me, acknowledging that this is an imperfect and necessarily partial interpretation. In addition, it is worth noting that the perspectives of the individuals may have changed as a result of their leaving the military, therefore their experiences and perspectives on those experiences may differ from those who are *currently* serving in these roles. As a final note on these issues, I have given pseudonyms to all of the individuals that I interviewed because of the contentiousness of what Reaper crews do. The majority of my interviewees requested pseudonyms and it made ethical sense to apply the same protection to all. For similar reasons, I did not record any of my interviews, the quotes that are provided in this book represent quotes that I carefully noted down during the interview rather than transcriptions, as such it is possible that they are imperfect recordings of precisely what was said, and any errors are my own.

The shape of the beast

The book is primarily concerned with how using the framework of Haunting can help to more adequately shed light on the ways that gendered discourses are destabilised and (re)inscribed through the use of novel military technologies, and the structure of the book reflects the way in which I make this argument. As with all research, I have had to draw arbitrary lines between some of the chapters, particularly the empirical chapters, which seem to imply a separation between their concerns. However, as becomes clear from the content of these chapters; lives, discourses, experiences, and ghosts do not necessarily acknowledge the boundaries we draw, and therefore you will find moments of cross-reference, places where stories that seem over and done with in one chapter re-emerge in relation to another discussion. The aim of this book is not to get rid of the ghosts that emerge through the prism of Haunting, but rather to engage them in discussion. This is not to imply that the conversations are ones that resolve, but I hope the chapters provide spaces where both scream and silence can be heard.

Chapter 1 considers how feminist scholarship has theorised military technologies and how this has been important to understanding warfare, looking more widely at military technologies and wider trends in the study of warfare. In particular, how this thinking has portrayed technology as 'gender neutral'. Then I sketch out feminist security studies interventions into the wider discussion of war, concentrating on how feminist theorising reveals the work that gender *does* in relation to technological developments in warfare. In this section I bring to the fore the gendered binaries at work and how these, specifically, are implicated in our understandings of developments in weaponry and warfare. I focus on the gendered discourses deployed when speaking about robotics in warfare, as a specific iteration of military technology, and then sift through these narratives at work in scholarship on the use of armed drones.

Chapter 2 introduces the *means* through which I argue that military technologies simultaneously destabilise and (re)inscribe military masculinity (and femininity). In this chapter I introduce my framework of Haunting and the wider 'spectral (re)turn' of which it is part. I outline the different ways that ghosts are useful to understanding personhood as complex (and why it might not fit neatly into either a category of masculine or feminine), how the in/(hyper)visibility of things affects their power and influence, how time is not linear (the past and the present and future are all implicated in what we think and feel now, and now, and now), and how (cob)webs of power are as much created out of fiction and myth as fact (and are as much made out of the small, particular and domestic as they are made out of the universal and grand). Utilising the figure of the ghost as a trope that represents the inherent instability of binary constructions (being present whilst simultaneously absent, being silent whilst screaming, being visible whilst being invisible) I outline why I add Cynthia Weber's queer logic(s) to the framework of Haunting and how that addition is useful to understanding the way that ideas of masculinity and femininity work in discourses of military technology.

The ghost of the warrior forms the focus of Chapter 3. This ghost reveals the 'traditional' gendered discourses at work in discussions of what it means to be a member of the military, what it means to be a warrior. Tracing the warrior's varied and multiple (non)lineage I situate 'traditional' narratives as problematic, revealing the ways in which arguments about masculinity, and more widely about gender in this context, are not as straightforward as they first appear. I draw out different ways of seeing gender in relation to bodies, weaponry, sex and sexuality, as well as arguments about connections between birth and death (and the attendant connections between femininity and masculinity) through different iterations of the warrior.

Chapters 4, 5, and 6 draw the ghost hunt of the warrior into the realm of British Reaper crews. The first of these, Chapter 4, addresses perhaps the most controversial element of the armed drones: the capacity to kill at a distance. Focused around the question of whether the warrior has a duty to kill and a duty to die, I trace the emergence of the ghost of the 'Other' and the ghost of cowardice. With the capacity (authority and legitimacy) to kill coded as masculine it appears that Reaper crews' masculinities are (re)inscribed by the capacity to launch missiles; however, as the ghosts of this chapter reveal, it is not that simple. The ghost of the 'Other' draws attention to the inability of Reaper crews to prevent the deaths of friendly forces on the ground, with lack of intelligence or procedural restraints rendering them impotent in the face of improvised explosive devices (IEDs) or surprise attacks. The ghost of cowardice reveals the feminising discourse of being situated at an extreme distance from the theatre of war, apparently not at risk and therefore unable to perform the heroic, physically risky role of the warrior. Through these ghosts, the chapter asks whether being a warrior means having a duty to kill and a duty to die in order to establish the necessary credentials for military masculinity.

Chapter 5 moves on from the focus on lethal strikes to the function of persistent surveillance in destabilising and (re)inscribing military masculinity in British Reaper crews. Persistent surveillance is the task that Reaper crews spend their majority of time undertaking. Through Haunting, this chapter engages with issues of in/(hyper)visibility asking questions about the gendered nature of the relationship between the watcher and the watched. What makes Reaper crews so interesting in answering these questions is that they are positioned as both watcher (empowered/masculine) and watched (disempowered/feminised). The ghost of omniscience reveals the instability of the relationship between the watcher and masculinity (as understood as emotionally distanced) and the ghost of the watchman reflects the ways in which the Reaper crews occupy the feminised position of 'the watched' and in this section I consider the implications for the crews of being 'the most supervised platform bar none'.[40]

The final chapter asks if it is possible to be a warrior and return to the home and the domestic daily. Bringing the book full circle and back to the focus on the quotes that begin this introduction, Chapter 6 investigates how the blurring of boundaries of the war front and the home hearth has implications for military masculinity. This chapter traces the way that 'going away' to war is a fundamental part of constructing the warrior identity, and how this haunts the crews whose commute does not take them (physically) to the dirty, dusty battlefields of Iraq or Afghanistan but rather to the clean, quiet environment of the Ground Control Station (GCS), replete with associations with the feminine. In addition, the lack of 'going away' to war impacts on the crews' access to 'decompression', an important part of maintaining a distinction between the spheres of war and the home. This chapter draws out the implications of the intrusion of war thinking into the domestic space, and also of the intrusion of domestic concerns into the battlefield. The final section of this chapter traces the way that the chronic fatigue experienced by the Reaper crews acts as a both a feminising (in comparison with the cyborgian body of the drone) and masculinising (in overcoming difficulty and discomfort) component of the crews' lives. This chapter draws attention to the creation of strange liminal zones of confusion in the lives of Reaper crews, disrupting time and space to produce a 'borderland' that reflects the queer logics at work in this and the other elements of the Reaper crews' lives (Anzaldúa, 2012).

Together, these chapters argue that the introduction of novel military technologies does not result in either the feminisation or hyper-masculinisation of the militaries that use them. Rather, this book demonstrates the way that gender works at a symbolic level in discourses surrounding military technologies so that these technologies serve to simultaneously destabilise *and* (re)inscribe military masculinity.

Notes

1 (Sauer and Schornig, 2012; Wexler, 2003; Morton, 1998; and Cohn, 1987 among others).
2 (Van Creveld, 1991; Coker, 2013; Singer, 2011; Dunn, 2013; Gray, 2013).
3 (for example Goldstein, 2003; Enloe, 1983; Sjoberg, 2014).
4 On feminisation see: (for example, Coker, 2002; Vallor, 2013; Van Creveld, 2013a; Royakkers and van Est, 2010; Asaro, 2013), on hyper-masculinisation see: (for example, Kunashakaran, 2016; Masters, 2005; Bayard de Volo, 2016; Holmqvist, 2013).
5 (Daggett, 2015; Wilcox, 2015).
6 (see also Sauer and Schornig, 2012; Asaro, 2013; Royakkers and van Est, 2010).
7 I capitalise Haunting when I am referring to the theoretical framework that I use in this book and use 'haunting' when using the term otherwise.
8 (Elshtain, 1995; Hooper, 2001; Cohn, 1993).
9 (Elshtain, 1995; Hooper, 1998; Peterson and Runyan, 2010).
10 (Baaz and Stern, 2009; Allsep, 2013; Cohn, 1998, 2000).
11 Implications (Holmqvist, 2013; Sylvester, 2013; McSorley, 2014).
12 And certainly 'more complicated than those of us who study it have usually granted' (Gordon, 2008, p. 7).
13 (Haraway, 1988; Shepherd, 2008).
14 (Singer, 2011; Birmingham Policy Commission, 2014; Rogers and Hill, 2014; Enemark, 2015).
15 British armed forces have never used *British* Predators but have been embedded with US crews who were operating *American* Predators (Martin, 2010). The Royal Air Force does have its own collection/fleet of *Reaper* drones – Italian forces have recently acquired armed drones but have not yet used their lethal capacity.
16 (Terkel, 2015; Ouma *et al.*, 2011).
17 (With the notable exceptions of Lee, 2013, 2018; Birmingham Policy Commission, 2014).
18 (BBC, 2015; Office of the Spokesperson, 2015; Martin, 2010; Wright, 2016).
19 (for example, Jenkings *et al.*, 2011; Stavrianakis, 2006).
20 A concept used by Gordon and introduced in detail in Chapter 2.
21 In terms of the number of armed drones, the number of strikes, the number of ways in which armed drone strikes are used and investment in the development of armed drone technologies.
22 (Gertler, 2012; Barnidge, 2012; Proxdynamics, n.d.; Whetham, 2013; Brooke-Holland, 2013; Robinson, 2013; Emmerson, 2014; Gogarty and Robinson, 2012).
23 (Chuter and Stevenson, 2018; General Atomics Aeronautical, n.d.; Gogarty and Robinson, 2012).
24 (see, for example, Murch, 2016; Fung, 2013; Bennett-Jones, 2014; Wright, 2013; Wolfgang, 2013; Gosztola, 2013).
25 (Woodward, 2004; Woodward and Jenkings, 2011; Woodward and Winter, 2007; Duncanson, 2013; Basham, 2013).
26 (for example, Chappelle *et al.*, 2014; Ouma *et al.*, 2011; Drew and Philipps, 2015; Fitzsimmons and Sangha, n.d.).
27 (Terkel, 2015; Cantwell, 2009).
28 (Kretzmer, 2005; Fang, 2015; Carvin, 2012; Enemark, 2015).
29 (Rogers and Hill, 2014; Heller, 2013).
30 (Williams, 2010; Alston and Shamsi, 2010; Hastings, 2012; Bowden, 2013; Gregory, 2011; Adams and Barrie, 2013; Kretmer, 2005).

31 (Ackerman, 2016; Shishkin *et al.*, 2015).
32 (Aslam, 2011; Taj, 2010; Kretzmer, 2005).
33 (Fang, 2015; Shane, 2015; Hudson *et al.*, 2011; Swift, 2012; Rogers and Hill, 2014; Dowd, 2013).
34 As reported by the MOD: www.raf.mod.uk/aircraft/mq-9a-reaper/ (accessed 19 December 2018).
35 (with updated information available at: www.gov.uk/government/news/update-air-strikes-against-daesh) https://dronewars.net/uk-drone-strike-list-2/ (as up to 19/12/2018).
36 https://dronewars.net/uk-drone-strike-list-2/ (accessed 19 December 2018).
37 However, when former UK Prime Minister David Cameron commanded the use of a drone strike against two British citizens (who had travelled to Syria apparently to plan terrorist attacks against the United Kingdom (Watt *et al.*, 2015)) this reflected some concerning similarities to US strikes, in that its targets were British citizens and is considered by some to represent 'mission creep' (MacAskill, 2015).
38 (Chris Cole, 2015b; Serle and Fielding-Smith, 2015).
39 (Ross and Ball, 2015; Shabibi and Watling, 2016; Amnesty International, 2018).
40 Interview with 'Dan'.

References

Ackerman, S. (2016) US drone strike kills key al-Shabaab leader in Somalia. *Guardian* [online], 4 January. Available at: www.theguardian.com/world/2016/apr/01/us-military-somalia-airstrike-al-shabaab-terrorist-targets (accessed: 4 May 2016).

Alexander, D. (2015) Air Force moves to ease stress on overworked U.S. drone pilots. *Reuters*, 16 January. Available at: http://in.reuters.com/article/2015/01/16/usa-defense-drones-idINL1N0UU35M20150116 (accessed: 7 March 2015).

Allsep, L. Michel (2013) The myth of the warrior: martial masculinity and the end of don't ask, don't tell. *Journal of Homosexuality*, 60 (2): 381–400.

Alston, P. and Shamsi, H. (2010) A killer above the law? *Guardian* [online], 8 February. Available at: www.theguardian.com/commentisfree/2010/feb/08/afghanistan-drones-defence-killing (accessed: 4 January 2015).

Amnesty International. (2018) Deadly assistance: The role of European states in US drone strikes. ACT 30/8151/2018 www.amnesty.org/download/Documents/ACT3081512018 ENGLISH.PDF (accessed: 19 December 2018).

Anderson, K. (2000) Ottawa Convention banning landmines, the role of international non-governmental organizations and the idea of international civil society. *European Journal of International Law*, 11: 91.

Anderson, K. and Waxman, M. (2012) *Law and ethics for robot soldiers*. ID 2046375. Rochester, NY: Social Science Research Network. Available at: http://papers.ssrn.com/abstract=2046375 (accessed: 13 March 2013).

Anzaldúa, G. (2012) *Borderlands: la frontera: the new Mestiza*. 4th edn. San Francisco, CA: Aunt Lute Books.

Arkin, R. (2015) The case for banning killer robots: counterpoint. *Communications of the ACM*, 58 (12): 46–47. doi: 10.1145/2835965.

Asaro, P.M. (2013) The labor of surveillance and bureaucratized killing: new subjectivities of military drone operators. *Social Semiotics*, 23 (2): 196–224. doi: 10.1080/10350330.2013.777591.

Aslam, M.W. (2011) A critical evaluation of American drone strikes in Pakistan: legality, legitimacy and prudence. *Critical Studies on Terrorism*, 4 (3): 313–329.

Auchter, J. (2014) *The politics of haunting and memory in international relations*. Interventions. London; New York: Routledge/Taylor & Francis Group.

Bal, M. (2010) Guest column: exhibition practices. *PMLA*, 125 (1): 9–23. doi: 10.1632/pmla.2010.125.1.9.

Balsamo, A.M. (1996) *Technologies of the gendered body: reading cyborg women*. Durham: Duke University Press.

Barnidge, R. (2012) A qualified defense of American drone attacks in northwest Pakistan under international humanitarian law. *Boston University International Law Journal*, 30 (3): 410–447.

Basham, V. (2013) *War, identity and the liberal state: everyday experiences of the geopolitical in the Armed Forces*. Interventions. 1st edn. London; New York: Routledge, Taylor & Francis Group.

Bayard de Volo, L. (2013) Unmanned? Drones and the revolution in gender-military affairs. *Conference Paper for 3rd European Conference on Politics and Gender*. Available at: www.ecpg-barcelona.com/sites/default/files/Ppr-Unmanned-ECPG.pdf.

Bennett-Jones, O. (2014) Drones or UAVs? The search for a more positive name. *BBC Magazine* [online], 2 February. Available at: www.bbc.co.uk/news/magazine-25979068 (accessed: 26 May 2015).

BBC. (2015) US to allow armed drone exports. Available at: www.bbc.co.uk/news/world-us-canada-31523207 (accessed: 29 June 2015).

The Birmingham Policy Commission. (2014) The security impact of drones. Available at: www.birmingham.ac.uk/research/impact/policy-commissions/remote-warfare/index.aspx

Black, B. (2013) *A rare insight into a day at work for a pilot of a drone*. Available at: www.bbc.co.uk/news/world-24522150 (accessed: 27 January 2015).

Blanco, M. del P. and Peeren, E. (eds) (2013) *The spectralities reader: ghosts and haunting in contemporary cultural theory*. New York: Bloomsbury Academic.

Bowden, M. (2013) The killing machines. *The Atlantic* [online], 14 August. Available at: www.theatlantic.com/magazine/archive/2013/09/the-killing-machines-how-to-think-about-drones/309434/?single_page=true

Boyle, M.J. (2013) The costs and consequences of drone warfare. *International Affairs*, 89 (1): 1–29.

Brooke-Holland, L. (2013) *Unmanned aerial vehicles (drones): an introduction*. SN06493. House of Commons Library.

Brooke-Holland, Louisa (2015) Overview of military drones used by the UK armed forces. Report number: 06493 House of Commons Library Pub 11 June 2015, Available at: http://researchbriefings.files.parliament.uk/documents/SN06493/SN06493.pdf (accessed: 1 July 2015).

Burkeman, O. (2009) Obama administration says goodbye to 'war on terror'. *Guardian*, 25 March 2009. Available at: www.theguardian.com/world/2009/mar/25/obama-war-terror-overseas-contingency-operations (accessed: 7 January 2019).

Calhoun, L. (2011) The end of military virtue. *Peace Review*, 23 (3): 377–386. doi: 10.1080/10402659.2011.596085.

Cantwell, H.R. (2009) Operators of Air Force unmanned aircraft systems: breaking paradigms. *Air and Space Power Journal*, 33 (2): 67–77.

Chapa, J.O. (2014) Remotely piloted aircraft and war in the public relations domain. *Air and Space Power Journal*, (September–October): 29–46.

Chappelle, W.L., McDonald, K.D., Prince, L., Goodman, T., Ry-Sannerud, B.N. and Thompson, W. (2014) Symptoms of psychological distress and post-traumatic stress disorder in United States Air Force 'drone' operators. *Military Medicine*, 179 (8S): 63–70.

Chuter, Andrew and Stevenson, Beth (2018) Britain chooses basing for Protector drone, even as acquisition details evolve. *Defense News*. Available at: www.defensenews. com/air/2018/07/13/britain-chooses-basing-for-protector-drone-even-as-acquisition-process-evolves/ (accessed: 19 December 2018).

Cohn, C. (1987) Sex and death in the rational world of defense intellectuals. *Signs*, 12 (4): 687–718.

Cohn, C. (1993) War, wimps, and women: talking gender, thinking war. In Cooke, M. and Woollacott, A. (eds) *Gendering war talk*. Princeton, NJ: Princeton University Press.

Cohn, C. (1998) Gays in the military: text and subtext. In Zalewski, M. (ed.) *The 'man question' in international relations*. Builder, CO: Westview Press.

Cohn, C. (2000) 'How can she claim equal rights when she doesn't have to do as many push-ups as I do?': the framing of men's opposition to women's equality in the military. *Men and Masculinities*, 3 (2): 131–151.

Cohn, C. and Ruddick, S. (2003) *A feminist ethical perspective on weapons of mass destruction*. Consortium on Gender, Security and Human Rights Working Paper No. 104. Available at: http://genderandsecurity.org/sites/default/files/carol_cohn_and_sara_ruddick_working_paper_104.pdf (accessed: 19 September 2014).

Coker, C. (2002) *Waging war without warriors?: the changing culture of military conflict*. Boulder, CO: Lynne Rienner Publishers.

Coker, C. (2007) *The warrior ethos: military culture and the war on terror*. LSE international studies series. London; New York: Routledge.

Coker, C. (2013) *Warrior geeks: how 21st century technology is changing the way we fight and think about war*. London: C. Hurst & Co. Publishers Ltd.

Cole, Chris (2015a) New figures for British air and drone strikes in Iraq. Drone Wars UK, Available at: http://dronewars.net/2015/05/15/new-figures-for-british-air-and-drone-strikes-in-iraq/ (accessed: 7 March 2019).

Cole, Chris (2015) Drones aren't just toys that cause a nuisance. They're still killing innocent people. *Guardian. Available at:* www.theguardian.com/commentisfree/2015/mar/20/drones-nuisance-killing-innocent-people (accessed: 4 June 2015).

Crane-Seeber, J.P. (2016) Sexy warriors: the politics and pleasures of submission to the state. *Critical Military Studies*, 2 (1–2): 41–55. doi: 10.1080/23337486.2016.1144402.

Creed, B. (1993) *The monstrous-feminine: film, feminism, psychoanalysis*. Popular fiction series. London; New York: Routledge.

Daggett, C. (2015) Drone disorientations: how 'unmanned' weapons queer the experience of killing in war. *International Feminist Journal of Politics*, 17 (3): 361–379. doi: 10.1080/14616742.2015.1075317.

Donnelly, S. (2005) Long-distance warriors. *TIME*, pp. 42–43.

Dowd, A.W. (2013) Drone wars: risks and warnings. *Parameters*, 42–43 (4–1): 7–16.

Doyle, J. (2013) Rise of the robots? Western unmanned air operations in Iraq and Afghanistan, 2001–2010. *Air Power Review*, 16 (2): 10–31.

Drew, C. and Philipps, D. (2015) As stress drives off drone operators, air force must cut flights. *New York Times* [online], 16 June. Available at: www.nytimes.com/2015/06/17/us/as-stress-drives-off-drone-operators-air-force-must-cut-flights.html?ref=us&_r=0 (accessed: 29 June 2015).

Duncanson, C. (2013) *Forces for good?* [online]. London: Palgrave Macmillan UK. Available at: http://link.springer.com/10.1057/9781137319425 (accessed: 27 June 2016).

Dunn, David H. (2013) Drones: disembodied aerial warfare and the unarticulated threat. *International Affairs*, 89 (5): 1237–1246.

Dworkin, A. and European Council on Foreign Relations (2013) *Drones and targeted killing: defining a European position*. London: ECFR.

Ehrenreich, B. (1998) *Blood Rites*. St Ives, Great Britain: Virago.

Elshtain, J.B. (1995) *Women and war*. University of Chicago Press ed. Chicago, IL: University of Chicago Press.

Emmerson, B. (2014) Report of the Special Rapporteur on the promotion and protection of human rights and fundamental freedoms while countering terrorism. 25th Session A/HRC/25/59. United Nations Human Rights Council.

Enemark, Christian (2015) *Armed drones and the ethics of war: military virtue in a post-heroic age*. New York: Routledge.

Enloe, C.H. (1983) *Does khaki become you?: the militarisation of women's lives*. London: Pluto Press.

Enloe, C.H. (1989) *Bananas, beaches & bases: making feminist sense of international politics*. London: Pandora.

Erikkson Baaz, Maria and Stern, Maria (2009) Why do soldiers rape? Masculinity, violence, and sexuality in the armed forces in the Congo (DRC). *International Studies Quarterly*, 53 (2): 495–518.

Fang, M. (2015) Nearly 90 percent of people killed in recent drone strikes were not the target. *Huffington Post* [online], 20 October. Available at: www.huffingtonpost.com/entry/civilian-deaths-drone-strikes_us_561fafe2e4b028dd7ea6c4ff (accessed: 4 May 2016).

Fitzsimmons, S. and Sangha, K. (n.d.) Killing in high definition: combat stress among operators of remotely piloted aircraft. *Technology*, 12: 289–292.

Freeman, C. (2015) Armchair killers: life as a drone pilot. *Telegraph*, 4 November. Available at: www.telegraph.co.uk/culture/film/film-news/11525499/good-kill-drone-pilot-true-story.html (accessed: 30 June 2015).

Fung, B. (2013) Why drone makers have declared war on the word 'drone'. *Washington Post* [online], 16 August. Available at: www.washingtonpost.com/blogs/the-switch/wp/2013/08/16/why-drone-makers-have-declared-war-on-the-word-drone/ (accessed: 26 May 2015).

Futter, A. (2011) NATO, ballistic missile defense and the future of US tactical nuclear weapons in Europe. *European Security*, 20 (4): 547–562. doi: 10.1080/09662839.2011.626404.

Gartzke, E. (2013) The myth of cyberwar: bringing war in cyberspace back down to earth. *International Security*, 38 (2): 41–73.

General Atomics Aeronautical (n.d.) Predator XP RPA. [online]. Available at: www.ga-asi.com/predator-xp (accessed: 26 May 2015).

Gertler, J. (2012) *U.S. unmanned aerial systems*. CRS Report for Congress 7–5700. Congressional Research Service

Gogarty, B. and Robinson, I. (2012) Unmanned vehicles: a (rebooted) history, background and current state of the art. *Journal of Law, Information & Science* [online], 21 (1). Available at: www.austlii.edu.au/au/journals/JlLawInfoSci/2012/2.html (accessed: 26 May 2015).

Goldstein, Joshua (2003) *War and gender: how gender shapes the war system and vice versa*. Cambridge: Cambridge University Press.

Gordon, A. (2008) *Ghostly matters: haunting and the sociological imagination*. New University of Minnesota Press ed. Minneapolis, MN: University of Minnesota Press.

Gordon, A. (2011) Some thoughts on haunting and futurity. *Borderlands*, 10 (2): 1–21.

Gosztola, Kevin, (2013) Don't use the 'D' word: they're 'UAVs' or 'RPAs' but definitely not 'drones'. *The Dissenter.* Available at: http://dissenter.firedoglake.com/2013/10/11/dont-use-the-word-drones-theyre-uavs/ (accessed: 26 May 2015).

Gray, H. (2016) Researching from the spaces in between? The politics of accountability in studying the British military. *Critical Military Studies*, 2 (1–2): 70–83. doi: 10.1080/23337486.2016.1127554.

Gray, Peter. (2013) The ethics of warfare, part 2: how do you define a combatant? *Birmingham Perspective.* Available at: www.birmingham.ac.uk/research/perspective/ethics-of-warfare-pt2.aspx (accessed: 16 February 2016).

Gregory, D. (2011) From a view to a kill: drones and late modern war. *Theory, Culture & Society*, 28 (7–8): 188–215.

Guerrina, R. and Wright, K.A.M. (2016) Gendering normative power Europe: lessons of the women, peace and security agenda. *International Affairs*, 92 (2): 293–312. doi: 10.1111/1468–2346.12555.

Haraway, D. (1991) A cyborg manifesto: science, technology and socialist feminism in the late twentieth century. In *Simians, cyborgs and women: the reinvention of nature.* New York: Routledge. pp. 149–181.

Haraway, D. (1997) *Modest-Witness@Second-Millennium.FemaleMan-Meets-OncoMouse: feminism and technoscience.* New York: Routledge.

Haraway, D. (1988) Situated knowledges: the science question in feminism and the privilege of partial perspective. *Feminist Studies*, 14 (3): 575.

Harris, V. (2015) Hauntology, archivy and banditry: an engagement with Derrida and Zapiro. *Critical Arts*, 29 (sup1): 13–27. doi: 10.1080/02560046.2015.1102239.

Hastings, M. (2012) The rise of the killer drones: how America goes to war in secret. *Rolling Stone* [online], 16 April, (1155). Available at: www.rollingstone.com/politics/news/the-rise-of-the-killer-drones-how-america-goes-to-war-in-secret-20120416 (accessed: 4 August 2015).

Heller, K.J. (2013) 'One hell of a killing machine': signature strikes and international law. *Journal of International Criminal Justice*, 11 (1): 89–119.

Holmqvist, C. (2013) Undoing war: war ontologies and the materiality of drone warfare. *Millennium – Journal of International Studies*, 41 (3): 535–552. doi: 10.1177/0305829813483350.

Hooper, Charlotte (1998) Multiple masculinities in international relations. In Zalewski, M. (ed.) *The 'man question' in international relations.* Boulder, CO: Westview Press.

Hooper, Charlotte (2001) *Manly states: masculinities, international relations, and gender politics.* New York: Colombia University Press.

House of Commons Defence Committee (2014a) Remote control: remotely piloted air systems – current and future UK use. Tenth Report of Session 2013–14 HC772, 11 March 2014.

House of Commons Defence Committee (2014b) Remote control: remotely piloted air systems – current and future UK use: Government Response to the Committee's Tenth Report of Session 2013–14, Sixth Special Report of Session 2014–15 HC 611, Published 29 July 2014.

Hudson, L., Owens, C.S. and Flannes, M. (2011) Drone warfare: blowback from the new American way of war. *Middle East Policy*, 18 (3): 122–132.

Jenkings, Neil, K., Woodward, Rachel, Williams, Alison, J., Rech, Matthew F., Murphy Ann, L. and Bos, Daniel; (2011) Military occupations: methodological approaches and the military-academy research nexus: military occupations. *Sociology Compass*, 5 (1): 37–51.

Kaurin, P. (2010) With fear and trembling: an ethical framework for non-lethal weapons. *Journal of Military Ethics*, 9 (1): 100–114. doi: 10.1080/15027570903523057.

Kontour, K. (2012) The governmentality of battlefield space: efficiency, proficiency, and masculine performativity. *Bulletin of Science, Technology & Society*, 32 (5): 353–360. doi: 10.1177/0270467612469067.

Kretzmer, D. (2005) Targeted killing of suspected terrorists: extra-judicial executions or legitimate means of defence? *European Journal of International Law*, 16 (2): 171–212.

Kunashakaran, S. (2016) Un(wo)manned aerial vehicles: an assessment of how unmanned aerial vehicles influence masculinity in the conflict arena. *Contemporary Security Policy*, 37 (1): 31–61. doi: 10.1080/13523260.2016.1154405.

Lash, J. (1995) *The hero: manhood and power*. Art and imagination series. New York: Thames and Hudson.

Lee, P. (2013) Rights, wrongs and drones: remote warfare, ethics and the challenge of just war reasoning. *Air Power Review*, 16 (3): 30–49.

Lee, P. (2015) *Truth wars: the politics of climate change, military intervention and financial crisis*. London: Palgrave Macmillan.

Lee, P. (2018) *Reaper force: the inside story of Britain's drone wars*. London: John Blake Publishing.

MacAskill, E. (2015) Drone killing of British citizens in Syria marks major departure for UK. *Guardian*, 9 July. Available at: www.theguardian.com/world/2015/sep/07/drone-british-citizens-syria-uk-david-cameron (accessed: 10 July 2016).

Manjikian, M. (2014) Becoming unmanned: the gendering of lethal autonomous warfare technology. *International Feminist Journal of Politics*, 16 (1): 48–65.

Martin, M.J. (2010) *Predator: the remote-control air war over Iraq and Afghanistan: a pilot's story*. Minneapolis, MN: Zenith Press.

Masters, C. (2005) Bodies of technology: cyborg soldiers and militarized masculinities. *International Feminist Journal of Politics*, 7 (1): 112–132. doi: 10.1080/1461674042000324718.

Mayer, J. (2009) The Predator war: what are the risks of the CIA's covert drone program? *The New Yorker*, 26 October.

McKinley, R.A., McIntire, L.K. and Funke, M.A. (2011) Operator selection for unmanned aerial systems: comparing video game players and pilots. *Aviation, Space, and Environmental Medicine*, 82 (6): 635–642. doi: 10.3357/ASEM.2958.2011.

McSorley, Kevin (2014) Towards an embodied sociology of war. *The Sociological Review*, 62: 107–128.

Morton, Jeffrey S. (1998) The legal status of laser weapons that blind. *Journal of Peace Research*, 35 (6): 697–705.

Mulrine, A. (2012) Drone pilots: why war is also hard for remote soldiers. *Christian Science Monitor*, 28 February. Available at: www.csmonitor.com/USA/Military/2012/0228/Drone-pilots-Why-war-is-also-hard-for-remote-soldiers.

Murch, L. (2016) Semantics, PR and the dreaded 'D' word. *Aerospace*, 43 (1): 30–31.

NATO (2003) *Cyberwar-Netwar*. Advanced Research Workshop on Cyberwar-Netwar: Security in the Information Age. IOS Press.

Office of the Spokesperson (2015) U.S. export policy for military unmanned aerial systems. Available at: www.state.gov/r/pa/prs/ps/2015/02/237541.htm (accessed: 29 June 2015).

Orbons, S. (2012) Are non-lethal weapons a viable military option to strengthen the hearts and minds approach in Afghanistan? *Defense & Security Analysis*, 28 (2): 114–130. doi: 10.1080/14751798.2012.678163.

Ouma, J.A., Chappelle, W.L. and Salinas, A. (2011) *Facets of occupational burnout among U.S. Air Force active duty and National Guard/Reserve MQ-1 Predator and MQ-9 Reaper operators*. 88ABW-2011–4485. Air Force Research Laboratory: Aerospace Medicine Education.

Peeren, E. (2014) *The spectral metaphor*. Palgrave Macmillan. Available at: www.palgraveconnect.com/doifinder/10.1057/9781137375858 (accessed: 28 April 2016).

Peterson, V.S. (ed.) (1992) *Gendered states: feminist (re)visions of international relations theory*. Gender and political theory. Boulder, CO: Lynne Rienner.

Peterson, V.S. and Runyan, A.S. (2010) *Global gender issues in the new millennium. Dilemmas in world politics*. 3rd edn. Boulder, CO: Westview Press.

Proxdynamics (n.d.) PD-100 Black Hornet. *Proxdynamics* [online]. Available at: www.proxdynamics.com/products/pd-100-black-hornet-prs (accessed: 26 May 2015).

Rayner, A. (1994) Cyborgs and replicants: on the boundaries. *Discourse: Studies in the Cultural Politics of Education*, 16 (3): 124–143.

Robinson, N. (2013) *Drone pilot. Cool military careers*. Ann Arbor, MI: Cherry Lake Pub.

Rogers, Ann and Hill, John (2014) *Unmanned: drone warfare and global security*. Ontario: Between the Lines.

Rogoway, T. (2015) Hollywood gives drone crews the American sniper treatment with good kill. *FoxTrot Alpha*. Available at: http://foxtrotalpha.jalopnik.com/hollywood-gives-drone-crews-the-American-sniper-treatme-1694560381.

Ross, Alice and Ball, James (2015) GCHQ documents raise fresh questions over UK complicity in US drone strikes. *Guardian*. Available at: www.theguardian.com/uk-news/2015/jun/24/gchq-documents-raise-fresh-questions-over-uk-complicity-in-us-drone-strikes (accessed: 17 May 2015).

Ross, A. and Woods, C. (2012) *Revealed: US and Britain launched 1,200 drone strikes in recent wars*. December. *The Bureau of Investigative Journalism*. Available at: www.thebureauinvestigates.com/stories/2012-12-04/revealed-us-and-britain-launched-1-200-drone-strikes-in-recent-wars (accessed: 16 April 2019).

Royakkers, L. and van Est, R. (2010) The cubicle warrior: the marionette of digitalized warfare. *Ethics and Information Technology*, 12 (3): 289–296. doi: 10.1007/s10676–010–9240–8.

Royal Air Force (2016) *Reaper MQ9A RPAS*. Available at: www.raf.mod.uk/equipment/reaper.cfm (accessed: 4 June 2016).

Sauer, F. and Schornig, N. (2012) Killer drones: the 'silver bullet' of democratic warfare? *Security Dialogue*, 43 (4): 363–380. doi: 10.1177/0967010612450207.

Serle, J. (2014) More than 2,400 dead as Obama's drone campaign marks five years. *Bureau of Investigative Journalism*, 23 January. Available at: www.thebureauinvestigates.com/2014/01/23/more-than-2400-dead-as-obamas-drone-campaign-marks-five-years/ (accessed: 17 September 2014).

Serle, Jack and Fielding-Smith, Abigail. (2015) Britain has flown 301 Reaper drones missions against ISIS in Iraq, firing at least 102 missiles, *The Bureau of Investigative Journalism*. Available at: www.thebureauinvestigates.com/2015/05/15/revealed-britain-has-flown-301-reaper-drone-missions-against-isis-in-iraq-firing-at-least-102-missiles/ (accessed: 5 May 2015).

Shabibi, Namir and Watling, Jack (2016) How the UK secretly helped direct lethal US drone strikes in Yemen. *Vice News*. Available at: https://news.vice.com/article/exclusive-how-the-uk-secretly-helped-direct-lethal-us-drone-strikes-in-yemen (accessed: 17 May 2015).

Shane, S. (2015) Drone strikes reveal uncomfortable truth: U.S. is often unsure about who will die. *New York Times* [online]. Available at: www.nytimes.com/2015/04/24/world/asia/drone-strikes-reveal-uncomfortable-truth-us-is-often-unsure-about-who-will-die.html?_r=0 (accessed: 22 October 2015).

Shepherd, L.J. (2006) Veiled references: constructions of gender in the Bush administration discourse on the attacks on Afghanistan post-9/11. *International Feminist Journal of Politics*, 8 (1): 19–41. doi: 10.1080/14616740500415425.

Shepherd, L.J. (2008) *Gender, violence and security: discourse as practice*. London; New York: Zed Books; distributed exclusively in the USA by Palgrave Macmillan.

Shishkin, P., Gross, J. and Abdulrahim, R. (2015) U.S. 'reasonably certain' drone strike killed 'Jihadi John'. *Wall Street Journal* [online], 13 November. Available at: www.wsj.com/articles/u-s-targets-jihadi-john-in-drone-strike-1447417423 (accessed: 4 May 2016).

Singer, P.W. (2011) *Wired for war: the robotics revolution and conflict in the 21st century*. New York: Penguin.

Sjoberg, Laura (2014) *Gender, war, and conflict*. Cambridge: Polity Press.

Stavrianakis, A. (2006) Call to arms: the university as a site of militarised capitalism and a site of struggle. *Millennium*, 35 (1): 139–154.

Steans, J. (2008) Telling stories about women and gender in the war on terror. *Global Society*, 22 (1): 159–176. doi: 10.1080/13600820701740795.

Swift, C. (2012) The drone blowback fallacy. *Foreign Affairs* [online], 7 January. Available at: www.foreignaffairs.com/articles/middle-east/2012-07-01/drone-blowback-fallacy (accessed: 4 May 2016).

Sylvester, Christine (2013) *War as experience: contributions from international relations and feminist analysis*. Milton Park, Abingdon, Oxon; New York: Routledge.

Taj, F. (2010) The year of the drone misinformation. *Small Wars & Insurgencies*, 21 (3): 529–535.

Terkel, William, (2015) Air force worried no one wants to be a drone pilot. *Huffington Post*. 16 January 2015, www.huffingtonpost.com/2015/01/16/drone-pilots_n_6488600.html (accessed: 16 March 2016).

The Economist (2014) Dilbert at war: the stressful lives of the 'Chair Force'. *The Economist*, 21 June.

Ussher, J.M. (2006) *Managing the monstrous feminine: regulating the reproductive body*. Women and psychology. 1st edn. London; New York: Routledge.

Vallor, S. (2013) *The future of military virtue: autonomous systems and the moral deskilling of the military*. Tallinn: NATO CCD COE Publications.

Van Creveld, Martin (1991) *Technology and war: from 2000 BC to the present*. 1st edn. New York: Free Press.

Van Creveld, M. (2013a) To wreck a military. *Small Wars Journal*. Available at: http://smallwarsjournal.com/jrnl/art/to-wreck-a-military (accessed: 13 October 2015).

Van Creveld, M. (2013b) *Wargames: from gladiators to gigabytes*. Cambridge; New York: Cambridge University Press.

Watt, Nicholas, Wintour, Patrick and Dodd, Vikram (2015) David Cameron faces scrutiny over drone strikes against Britons in Syria. *Guardian*. Available at: www.theguardian.com/world/2015/sep/07/david-cameron-justifies-drone-strikes-in-syria-against-britons-fighting-for-isis (accessed: 3 February 2016).

Weber, C. (2014) From queer to queer IR. *International Studies Review*, 16 (4): 596–601. doi: 10.1111/misr.12160.

Weber, C. (2016) *Queer international relations: sovereignty, sexuality and the will to knowledge*. Oxford studies in gender and international relations. New York: Oxford University Press.

Wexler, Lesley (2003) International deployment of shame, second-best responses, and norm entrepreneurship: the campaign to ban landmines and the landmine ban treaty. *Arizona Journal of International and Comparative Law*, 20: 561.

Whetham, D. (2013) Killer drones: the moral ups and downs. *The RUSI Journal*, 158 (3): 22–32.

Wilcox, Lauren, (2015) Drone warfare and the making of bodies out of place. *Critical Studies on Security*, 3 (1): 127–131.

Williams, B.G. (2010) The CIA's covert Predator drone war in Pakistan, 2004–2010: the history of an assassination campaign. *Studies in Conflict & Terrorism*, 33 (10): 871–892.

Wolfgang, B. (2013) Drone industry gives journalists not-so-subtle hint – don't use the word 'drones'. *Washington Times* [online], 14 August. Available at: www.washington times.com/news/2013/aug/14/drone-industry-journalists-dont-use-word-drones/ (accessed: 26 May 2015).

Woodward, R. (2004) Discourses of gender in the contemporary British Army. *Armed Forces & Society*, 30 (2): 279–301.

Woodward, R. and Jenkings, N.K. (2011) Military identities in the situated accounts of British military personnel. *Sociology*, 45 (2): 252–268.

Woodward, R. and Winter, T. (2007) *Sexing the soldier: the politics of gender and the contemporary British Army. Transformations, thinking through feminism*. London; New York: Routledge.

van Woudenberg, N. (2008) The long and winding road towards an instrument on cluster munitions. *Journal of Conflict and Security Law*, 12 (3): 447–483. doi: 10.1093/jcsl/krn007.

Wright, L. (2016) Armed drones: should the Canadian military use the controversial weapons? *CBC News* [online], 9 March. Available at: www.cbc.ca/news/canada/armed-drones-canadian-military-pros-and-cons-1.3482212 (accessed: 10 July 2016).

1   Theorising military technologies

The literature on military technologies is almost as diverse (or perhaps even more so) as the technologies themselves. Outside of the realm of the scientific, engineering and technical papers there is also a wealth of scholarship that addresses issues of importance to social science – the legal, ethical, social, psychological and cultural implications of the use of different kinds of technologies in warfare. One element that has often been overlooked, but which is of critical importance, is the work by feminist scholars on this issue. Whilst traditional security scholars have tended to view debates on gender as outside the remit of academic research into military technologies, feminist security scholars have definitively illustrated the myriad of ways and reasons that this is not the case. Instead, as feminist scholarship has indicated, 'gendering' research into the use of various technologies in warfare is fundamental to understanding the implications of their use. Beyond including gender as a 'variable' and 'adding women', feminist research has sketched out the co-constitution of gender and warfare (and by extension the technologies used in these contexts). By doing so, it becomes clear that it is not only *useful* to utilise a gendered lens for these topics but *essential* if we are to fully understand how military technologies function in the social, human context of war.

Existing interjections on this topic have explored the ways in which drone warfare can be constructed as a feminising influence on Western militaries (for example Coker, 2002; Vallor, 2013; Van Creveld, 2013; Royakkers and van Est, 2010; Asaro, 2013), and understood as representing their hypermasculinisation as a result of techno-fetishisation (for example, Kunashakaran, 2016; Masters, 2005; Bayard de Volo, 2016; Holmqvist, 2013). In addition, drone warfare has been explored as a queer phenomenon in that it disrupts the heteronormative structures that organise traditional narratives of warfare (Daggett, 2015; Wilcox, 2017). This latter approach reveals the instabilities, contradictions, and inherent 'queerness' within those structures. Daggett (2015) uses queer phenomenology specifically to examine the way drone warfare disrupts the 'distance-intimacy' and 'home-combat' spatio-temporal axes. Daggett's perspective offers an important insight into the value of recognising the fluidity/instability of gendered binaries, noting that 'warrior archetypes that order the act of killing … [are] defined against both the feminine and the queer' (Daggett, 2015, p. 362).

However, whilst Daggett's analysis 'can be read as concerning the maintenance of the heroic soldier myth and the production of martial violence as "combat"' (Millar and Tidy, 2017, pp. 149, 154), there needs to be a stronger (but also queer-inspired) engagement with the concept of the warrior in and of itself and how this is historically/socially/culturally situated. Therefore, what becomes clear is that there is a space in the literature for a framework which can do two things: First, allow us to take gender seriously in an analysis of military technologies and the social/human activity of warfare; and second, *simultaneously* enable a problematisation/destabilisation of the gendered binaries and dichotomous thinking that emerges from discussions of the co-constitution of gender and war. In this chapter, then, I sketch out the moves through the existing academic scholarship that brought me to the position of searching for such a framework, before the following chapter outlines the framework itself.

This chapter begins by describing some of the diverse work on the impact of technological developments on the conduct of warfare. Understanding warfare and technology as having a reciprocal relationship, I draw attention to the specific area of robots in war and their impact on the 'humanity' of war. I follow this with an illustration of the importance of considering the work that gender *does* in warfare, sketching out the important contribution(s) that feminist security scholars have made to the study of war. In addition to shining a light on the spaces and places where women have stories to tell about war and pointing out the importance of gendering *men* in warfare, this chapter outlines the importance of acknowledging the impact of gendered discourses in thinking about military technologies. I then pull together these various threads to consider how feminist security scholars currently address the issues raised by the use of armed drones as one iteration of the 'robotics' in war 'revolution', and have drawn upon (both explicitly and implicitly) the dualisms of gendered discourses. Through these discourses, I point to the need for a framework that is capable of embracing the complexity of these gendered narratives in ways that have not previously been adopted.

Military technologies

The development of technology has been a key component in warfare since the first projectiles were thrown (Braudy, 2005; Goldstein, 2001). As technical and scientific developments have enabled new feats of engineering the world has seen the introduction of slingshots, cross bows and long bows, guns, canons and missiles, as well as battleships, fighter jets, aircraft carriers, and armoured ground vehicles (to name but a few) (Rogers, 2011; Cook, 2000; Singer, 2011; Coker, 2013; Lambeth, 1997; Kaag and Kaufman, 2009). Beyond the importance of these developments from a *technological* perspective, there are also important ethical, legal, and human concerns that have emerged alongside the machinery (Gray, 2008; Geiβ, 2011; Casey-Maslen, 2014; Kaurin, 2010). Technology affects and changes the way that war is undertaken, changes the way that people engage in fighting, changes how people kill and are killed, changes how deaths are articulated in popular discourses, and changes how those individuals tasked with the activities of war operate and are perceived. Therefore understanding how technology is implicated in these questions is essential to understanding the much wider concern of warfare as a social, political and cultural phenomenon (Sylvester, 2013; McSorley, 2013).

The debates in this literature are grouped around specific concerns, either historical development (looking for wider trends), future-casting (sometimes accompanied by dire warnings about impending apocalypse), and grouped topics such as the ethics of the use of specific technologies in war. As commentators like Geoffrey Jensen and Ian Clark have noted, there has been a reluctance amongst military scholars to engage with the human-technology interface in war 'perhaps in part because they could not bear to turn their backs on the sort of military thinking that had dominated the more "heroic" times in which they felt most comfortable' (Jensen, 2001, p. 3; Clark, 2015). Therefore in some instances the most pertinent engagements with the issues came from social scientists and political thinkers like Friedrich Engels, Emile Durkheim, and Max Weber, who pointed to the interaction between, for example, the industrialisation of domestic society and the creation of war technologies (Engels, 1976; Jensen, 2001). For those military thinkers who did wrestle with the issue of technological development there was (and remains) disagreement over the degree to which material capabilities are more or less important than elements such as nationalism, leadership, and morale (compare, for example, the perspectives of Jomini, 2005; Engels, 1976; and Svechin and Lee, 2004). Liddell Hart's statement that ' "automatic warfare" … blows away romantic vapourings about the heroic values of war' is echoed in Christopher Coker's lament that 'For the true warrior, violence is existential … death had meaning. So too did honour, courage, and loyalty, all of which gave life meaning too… But by 1970… the warrior tradition was dead' (Liddell Hart, 1980, p. 33; Coker, 2002, p. 7).

These and other works, therefore, act as reminders that military technologies are not created in a vacuum, but rather reflect the contexts in which they are developed and the weight of political, social and cultural history (for example, McNeill, 1993; Van Creveld, 1991). For those concerned with the historical development of military technologies, there is a substantial literature devoted to discussing weaponry: the bow and arrow (Rogers, 2011), the gun (Reichmann, 1945), landmines (Anderson, 2000; Short, 1999), cluster munitions (Rappert and Moyes, 2009; van Woudenberg, 2008), nuclear weapons (Futter, 2011; Baylis and O'Neill, 2000), and, more recently, the advent of cyberwar (Gartzke, 2013; NATO, 2003), and the use of robots in warfare (Singer, 2011; Sparrow, 2007; IKV Pax Christi, 2011). In addition to discussing the specific conditions in which certain weapons developed, there is also significant interest in overarching trends and the way that this interweaves with certain political and social contexts: for example the First World War (Newpower, 2006), the Cold War (Oldenziel and Zachmann, 2009; Mahnken, 2010), and the peculiarity of the way in which the US and the UK harness technology in warfare (Brigety, 2007; Travers, 1992).

The relationship between war and technological development is complicated and co-constituting. Much time is spent by military scholars and practitioners attempting to predict both new technologies and new challenges, and how the new technologies will enable/prevent appropriate responses to the resulting new challenges, and how *these* challenges might be overcome by *other* new technologies and so on (Echevarria, 2007; McMaster, 2008; Barnaby, 1984; Beason, 2005; Coker, 2004). The need for better protection of troops and 'better' ways of killing the enemy are fed back into research and development, whilst at the same time developments in the technological sphere affect the way and means through which warfare is conducted. Referred to as 'revolutions in military affairs'[1] the impact of certain technologies in war have been heralded as changing the shape of warfare in particularly significant ways (Parker, 1996; Rogers, 1995). Different writers have indicated that different technologies have had effects that are more or less 'revolutionary'. For example, Max Boot (2007) identified four technological developments that have fundamentally changed the way that war is conducted: the advent of gunpowder, the first industrial revolution (with the advent of steam power), the second industrial revolution (with the advent of torpedoes and bombs), and the information revolution (with the advent of network-centric warfare, focusing on the impact of the computer). However, Boot's list is not considered definitive and some authors have indicated, for example, that nuclear weapons should be considered 'revolutionary' (Immerman and Goedde, 2013; Pick *et al.*, 1969; Neuneck, 2008) along with biological weapons (Martin, 2002; Galamas, 2008), and others still have looked back into history to argue for the revolutionary impact of the longbow (Neuneck, 2008; Rogers, 2011).

Despite this range of opinions, there is general agreement that technology and warfare have interacted (and continue to do so) in interesting and important ways. As such, Boot attempts to steer a route between technological determinism and not acknowledging the impact of changes in technology arguing, 'No technical advance by itself made a revolution, it was how people responded to technology that produced seismic shifts in warfare' (2007, p. 27). This statement is echoed in P.W. Singer's observation that

> World War I proved to be an odd, tragic mix of outmoded generalship combined with deadly new technologies. From the machine gun and radio to the aeroplane and tank, transformational weapons were introduced in the war, but the generals could not figure out how to use them.
>
> (Singer, 2011, p. 46)

It is therefore important, in a reverse situation of the earlier military scholars who eschewed detailing the impact of technological changes in preference for maintaining the myth of heroic leadership, not to let the technology itself overshadow the very human components of war. It is important not to forget that no matter whether it is by an arrow, a bullet, a missile, or a bomb; a soldier killed by any one of these is still dead. And that regardless of which of the above is deployed it is deployed by a human – whether by pressing a button, pulling a trigger, or launching a spear by hand.[2] Therefore this book engages with the phenomenon of military technology whilst trying to remain cognisant of the fact that war is fundamentally a human 'bodily and emotional experience', a 'politics of injury' based around the injuring and killing of other human beings (Sylvester, 2013, p. 1; see also Scarry, 1987; McSorley, 2013).

Whilst Boot and others reflect backwards on the historical developments in warfare in order to trace wider trends, this book is concerned with the specific implications of the introduction of robotics into warfare. I focus on this area because it is particularly salient and timely given the recent and rapid increase in the number and types of robots utilised in warfare (for example, Singer, 2011; Sparrow, 2007; *The Economist*, 2012). The academic literature on the introduction of robots to the battlefield has ranged widely from considerations of ethics, legality, autonomy, strategic effectiveness, and political implications (see Asaro, 2008; Pagallo, 2011; Danielson, 2011; Roff, 2014; Stone, 2004; Simpson, 2011). In discussions replete with references to *The Terminator* (1984) and *The Matrix* (1999) amongst other sci-fi and dystopian movies, commentators herald the era of the robot with a mixture of glee and terror (Sparrow, 2007; *The Economist*, 2012; Matthew, 2015). Many of these commentators identify the 'non-humanness' of robots as problematic in war: some argue that robots will be unable to act ethically and morally because of their 'non-humanness', whereas some argue that this means that they could be *more* ethical in their actions (lacking the human emotions of rage, jealousy and desire for revenge) (Sparrow, 2007; Asaro, 2008; IKV Pax Christi, 2011; Evangelista *et al.*, 2014). Aaron Johnson and Sidney Axinn (2013), amongst others, argue that robots should never be allowed to kill because this is a violation of human dignity (because the robot is not human) and therefore a violation of the laws relating to warfare. This apparent reduction in human *agency* in war is one that has galvanised the campaign against killer robots, which focuses on the future development of autonomous robots with the capacity to undertake lethal strikes (Sparrow, 2016; Sharkey, 2011, 2008). This book does not engage with the thorny and predictive issues regarding robotic autonomy. That topic is several books on its own. Rather it looks at the issues of the use of robots, *as they currently are*, in warfare, with a focus on a specific kind of military robot that is the drone.

Robots have been introduced in a range of different military roles, from ground vehicles that diffuse improvised explosive devices (IEDs), to underwater surveillance robots (Singer, 2011). However, the area of warfare that has seen by far the largest explosion of 'robots', is the area of unmanned or remotely piloted aerial vehicles, also called drones. Understood by Boot (2007) as part of information revolution, Singer argues for the specific significance of drones, describing them as 'the most important weapons development since the atomic bomb' (2011, p. 10; see also Coker, 2013). However, whilst the topic of drones has sparked massive debate and garnered a rapidly growing body of literature of its own, the *human* interaction with drones is one area that is currently under-investigated. If the problem with robots in warfare is, at least in part, because of their 'non-humanness' then it behoves us to look carefully at the areas where the humans *are* involved with the use of military robots. Whilst the discussion on the use of military technologies has resulted in debate across the academic spectrum from law to ethics, to politics and cultural studies, as well as engineering and computer science; this book approaches the issues from the perspectives of security studies as part of the wider academic discipline of International Relations. Specifically, this book speaks from and to the scholars that comprise feminist security studies. This is because, when we speak about human interactions with technology, the humans that we encompass in those statements are always already marked with a criss-crossing of identifiers that affect how people can and do behave. One of the most powerful of these identifiers is that of gender, and it is this that forms the primary concern for feminist scholars. The following section then goes on to outline the importance of gendering security studies more generally before moving on to sketching out the work done by feminist security scholars on military technologies in particular.

Feminist interjections

Military scholars and the sub-discipline of security studies has traditionally been concerned with the conduct of war, with strategic decisions made by leaders and with the conditions of the international system (Barkawi, 2011; Burgess, 2010; Hansen, 2009; Peoples, 2010; Williams, 2008). However, feminist theorists have pointed out that this conceptualisation of security is limited and ignores the important human component of security studies (Tickner, 1992). As the UN *Human development report* points out

> The concept of security has for too long been interpreted narrowly: as security of territory from external aggression.... Forgotten were the legitimate concerns of ordinary people who sought security in their daily lives.... For most people, a feeling of insecurity arises more from worries about daily life than from the dread of a cataclysmic world event....
> (United Nations Development Programme, 1994, pp. 22–23)

Therefore, taking the human seriously in terms of security topics means taking seriously the social and cultural structures that influence individual action and restrict agency. One of these structures, and the one that is of primary concern to feminist scholars is gender. Gender is integral to international relations and security because 'there is a structure to the international system and that structure cannot be understood without gender analysis' (Sjoberg, 2012, p. 3). From a feminist perspective, gender can be understood in multiple ways. It is something that can be understood as marking individual bodies, as something that affects how individuals and groups behave, and an analytical category (Peterson and Runyan, 2010). Thinking about gender in this way means that we can usefully apply a gender lens to problems at various different levels of international politics and security studies to help understand conceptualisations of International Relations, as well as thinking about how to solve security problems and promote 'positive change in the security realm' (Sjoberg, 2011, p. 600; Steans, 2013).[3]

Too often, mainstream security studies has expected gender theorists to provide a single 'gender' or 'feminist' perspective on a topic of interest, Marysia Zalewski, for example, sighs over the question 'Well, what is the Feminist Perspective on Bosnia?' (1995)[4] and Jill Steans notes that

> mainstream scholars assumed that standpoint theorists worked with settled and rather essentialist notions of gender, opening up the way for the introduction of a 'gender variable', rather than more postpositivist and gender-destabilising perspectives.
>
> (Steans, 2003, p. 437)

As noted in the introduction, I situate myself amongst the scholarship as a poststructural feminist. Therefore, the perspective of this book is one that aims to de-couple male/female from masculine/feminine, adopting a performative (after Butler, 1999), non-essentialist understanding of what gender *is* and how it functions.

Historically there has also been a tendency to minimise the scope of what feminist's study to 'women's issues' (Tickner, 1997; Sjoberg, 2012) and, even after years of debate, feminist scholars note that traditional security scholars[5] have been reluctant to embrace gender analyses because there is a perception that 'gender' has little to do with '"real-world" problems' (Tickner, 1997, p. 612). This perception occurs partly because some[6] traditional security scholars believe two things: that the world of 'real' security has little to do with women, and also because they believe that 'gender' only refers to women, both of which (as I go on to illustrate) are incorrect (Tickner, 1997; Zalewski, 1993; Weber, 1994; Carver, 1996).[7] As such, Jill Steans claims that

> At best mainstream scholars have engaged selectively with feminist IR, ignoring or even disparaging the work of scholars who work with unsettled notions of gender and gendered subjectivities, while selectively engaging with those scholars who seemingly worked with stable and unproblematic gender categories.
>
> (Steans, 2003, p. 430)

Such an approach limits a feminist analysis to talking about the differences between men and women (as stable and biologically defined categories), ignoring the impact of the social construction of masculinity and femininity (and their mutability). Similarly, Christine Sylvester also talks about this when she states 'IR is implicitly wedded to an unacknowledged and seemingly commonplace principle that international relations is the proper homestead or place for people called men', and yet women *are* involved in all parts of international relations, as the following section goes on to illustrate (1994, p. 4).

Where are the women?

Feminist scholars, like Cynthia Enloe (2000), have pointed out that women are present in much of international relations, but have simply been ignored. Utilising Enloe's question, 'Where are the women?', others have contributed to seeing women as part of the story of international relations by drawing out their distinctive stories and experiences (Moon, 1999; Goldstein, 2001; Smith, 1993). As V. Spike Peterson and Anne Sisson Runyan have pointed out, women have had many roles in international security, but these are obscured because women's roles have tended to be in the background as wives or carers, or in informal organisations such as non-governmental organisations (NGOs) rather than in governments (2010, p. 9). Additionally, women have been victims of the violence that occurs as a result of war and instability between states (the traditional remit of security studies) and it has been important to feminist scholars to provide a platform for enabling these marginalised voices to be heard, for women's stories to become part of the discourses that are heard about war, and what it means to be a part of violent conflict (Pettman, 1996; Afshar and Eade, 2004; D'Costa *et al.*, 2006; Enloe, 1993). Putting women back into security studies is not just about women's victimhood, but is also important in highlighting and talking about women's agency: 'Women are capturing hostages, engaging in suicide bombings, hijacking aeroplanes, and abusing prisoners' and it is necessary to find ways of speaking about and making sense of these phenomena (Sjoberg, 2007, p. 1; see also Shepherd, 2010). For example, when women engage as combatants; be that as soldiers (for example Baaz and Stern, 2011; Kennedy–Pipe, 2000; Just, 2006; Kronsell, 2012; MacKenzie, 2012; Stachowitsch, 2012), insurgents/guerrillas (Enloe, 1993; Sjoberg, 2010b) or terrorists (Gronnvoll and McCauliff, 2013; Rajan, 2011; Schweitzer and Merkaz Yafeh le-mehkarim astrategiyim, 2006; Marway, 2011).

Women are, and always have been, heavily involved in the key topics of security studies, both as victims and agents of violent conflict. Feminist scholars have illuminated women's lives and experiences in ways that have helped to paint a more complete picture of the subject of war, sensitive to the issue of gender through the inclusion of women. However, this is only part of the story. A 'gender' analysis within security studies is not *just* about putting women back in the picture.

Gendering men

As Terrell Carver notes, 'Gender is not a synonym for women' (Carver, 1996; see also Zalewski and Parpart, 1998).[8] Indeed it is very necessary to do a feminist gender analysis of the lives and experiences of men *as men* (Enloe, 2013). As such, doing a gender analysis means paying attention to the way that the power of gender infuses the lives of *both* men *and* women. It means revealing the disconnect between masculinity and men and femininity and women; and highlighting the ways that these socially constructed ideas of what it means to be a man or a woman are not unchanging or immutable but something that can and does change across time and social/cultural context (Peterson and Runyan, 2010). Many feminists now subscribe to the idea that gender is socially constructed. To argue that gender is socially constructed is to argue that there is 'no "essence" to gender', that there is nothing inherent in male or female biology that connects with masculinity or femininity (Shepherd, 2008a; 2010). As such, this book follows Sylvester's perspective on the social construction of gender:

> When speaking of 'men' and 'women', it is important to establish from the outset that I do not pose these gender categories as permanent, immutable, determinant, and essential. Rather, I see 'men' and 'women' as socially constructed subject statuses that emerge from a politicization of slightly different anatomies in ways that support grand divisions of labour, traits, places, and power. By 'socially constructed' I mean that men and women are the stories that have been told about 'men' and 'women'....
>
> (1994, p. 4)

Therefore, being a 'man' or a 'woman', is less about the body that you have and more about the 'stories' that are told that enable us to make sense of the categories of 'man' and 'woman', and 'male' and 'female', and 'masculine' and 'feminine'. Gender is constructed in language, in how men and women are spoken about, and this construction is profoundly political because it polices the ways that individuals are allowed to behave in a manner that invites violence in the case of transgression (Zalewski, 2000; Butler, 2011). From this perspective human subjectivity is something which we 'do', gender is something we must produce through repetitious actions (Peterson and Runyan, 2010; Butler, 1999). Gender then becomes, not the markings on the body but 'socially learned behaviours, repeated performances, and idealized expectations that are associated with and distinguish between the proscribed gender roles of masculinity and femininity'; produced by repetition of certain actions, by what is said, and by what is done (Peterson and Runyan, 2010, p. 2; Butler, 1999). In this way masculinity and femininity become discourses of gender.

In this context discourses are discursive practices that 'delineat[e] the terms of intelligibility whereby a particular reality can be "known" and acted upon' (Doty, 1996, p. 6). The 'stories' that Sylvester refers to are discourses, not just because they are stories that are spoken or written, but because they are 'stories' that are reproduced through action, that acquire meaning through repetition, and through practice as much as from the bodies that they are applied to (Sylvester, 1994). By de-coupling the concept of gender from biological sex we can argue that men 'acting as' men are not doing so because they are 'inherently' 'men' or that there is some innate 'maleness'. Similarly, women are not 'naturally' 'women' or channelling a biological 'femininity'. Taking this perspective disrupts the idea that men have some 'essence' that makes them better soldiers, leaders, or politicians; opening up the scope for women to be involved in these roles and making women's involvement in these roles intelligible (Zalewski and Parpart, 1998; Peterson and Runyan, 2010; Shepherd, 2008a; Sjoberg, 2007).

Gender and military technologies

Narratives of masculinity and femininity are reflected in discourses of military technologies and feminist and gender theorists have engaged directly with ideas, symbols, and discourses of masculinity, femininity, and sexuality in relation to military technologies, particularly weapons. Carol Cohn's work on US nuclear defence intellectuals is perhaps the most well-known and explicit in its findings. In *Sex and death in the world of rational defense intellectuals* (1987) Cohn sketches a range of ways that gendered themes appear. She notes, for example, the use of 'techno-strategic language' as a means of distancing the topic of discussion from the true horror of the use of a nuclear weapon, outlining the way 'rationality' as a masculine attribute is preferred by the defence intellectuals to acknowledging the possible outcomes in emotional language (which is coded as feminine).[9] These discussions are also wreathed in references to sex and sexuality. Cohn charts references to missiles 'releasing 70 to 80 percent of our megatonnage in one orgasmic whump', that are also referred to as 'penetrators' alongside discussions of the similarity between the 'loss of virginity' and the development of the first nuclear weapons, and with the identification of functional bombs as boys and the duds as girls, and on and surprisingly on (Cohn, 1987, p. 693).

Core themes in Cohn's work include the connection between masculinity and death as the inverse of the connection between the feminine and the capacity to give birth; the implicit references to sex and sexuality in discourses on dominance, protection, and watching; and linking the power of possessing a weapon with masculinity. These themes are also picked up by other feminist and gender scholars through a range of different means. Where Cohn focuses on the work gender is doing for the way that weapons are *talked* about, Joshua Goldstein (2001) argues for a connection between the way weapons are *used* and gender pointing to the phallic symbolism between the thrust of a sword or spear and the penetration of a body by the penis. Expanding on this idea, Goldstein draws on the work of Dave Grossman who noted that 'carrying a gun was like having a permanent hard-on' (Goldstein, 2001, p. 349; Grossman, 2009, Kindle location 2333). The link between the gun and 'having a permanent hard-on' may be construed partly as a result of the similarity in shape of the muzzle of the gun and the shape of an erect penis, but also because of the connection between penetration and power (Braudy, 2005). One of the (many) reasons that rape is a weapon of war is because it represents an intimate form of dominance. And whilst I strongly disagree with the statement, Arthur Brittan's argument that a 'man is only a man in so far as he is capable of using his penis as an instrument of power' is illustrative of this perspective (1989, pp. 46–47). Similarly, the connection between masculinity and sex and death, as noted by Cohn (1987), is reflected in other research that connects the process of killing with orgasm or loss of virginity. Attacking is described as 'the most exciting thing since getting my leg across' and the 'enormous pleasure and satisfaction' of a stream of bullets rushing out of the machine gun is likened to 'the orgasmic discharge' (Grossman, 2009, Kindle locations 2334–2346).

Looking beyond the advent of the machine gun and the nuclear weapon, feminist scholars have engaged with the gendering of new and novel technologies. The use of robotics in warfare, which forms the focus of this book, is an area of growing interest within feminist security studies. The introduction of robotics adds a new 'actor' into the dyad of 'just warrior' and 'beautiful soul' (Elshtain, 1995), challenging traditional narratives regarding who is the protector and who is protected. Indeed, Cristina Masters argues that 'the constitution of the cyborg solider is a radical rearticulation of subjectivity', because the technology has the capacity to displace 'men' as the 'subject capable of the discursive transcendence of embodiment' with implications for how wars are conducted, against whom and with what outcomes (2010, p. 4)

Perhaps the most well-known exploration of this issue is Donna Haraway's (1991) *Cyborg Manifesto*. There is not space here to unpack Haraway's work in all its creative detail, nor to engage in a complete review of the various debates that it has spawned (see, for example, Hamilton, 1997; Rayner, 1994; Gray, 1995). However, the figure of the cyborg is an important part of feminist thinking about technology, including military technologies. Haraway's cyborg is 'a cybernetic organism, a hybrid of machine and organism, a creature of social reality as well as a creature of fiction', a blend of human and machine, and of social and scientific (1991, p. 149). This blended entity, Haraway argues, has liberatory promise: the potential to disrupt gendered and heterosexed binaries, acting as 'an imaginative resource' for a 'post-gender world' (1991, pp. 149–150).

Where Donna Haraway heralded the emancipatory potential of the cyborg for taking us beyond the masculine/feminine binary, other authors have expressed concerns. In direct response to Haraway, Mary Manjikian (2014) argues that whilst we have yet to fully explore and understand the implications of battlefield robotics, it is unlikely that the sedimented way in which gender plays out in warfare will be easily overturned. There is the potential for transformation, but we should not necessarily expect this to result in post-gender betterment, as such she argues:

> The old dyad of protector/protected could be transformed when both men and women are protected by robots. If we accept Butler's (1990) argument that gender is performative, then the warfighter (whether male or female) might be coded as female, since he or she is now a passive agent requiring protection which is actively provided by a machine....
>
> (Manjikian, 2014, p. 50)

What Manjikian is arguing here is that, even if the warfighter is no longer a 'man', the *role* of warfighter will likely still be constructed as the masculine role. Rather than disrupting the distinction that situates the warfighter as masculine and the protected as feminine, the robot soldier will serve to re-sediment that distinction, albeit with a different actor in each of those positions.

The key type of military robot, and the focus of this book, is the drone. As outlined in the introduction chapter, this development is highly contentious and, recently, feminist scholars have made important interjections to thinking about the function of gender in relation to this specific military technology. Lauren Bayard de Volo argues that 'the new weaponry renders courage and physical strength nearly irrelevant, [so that] the traditionally valorised masculine attributes associated with heroism are eclipsed' with important implications for how gender and war interact (2013, p. 23). The implication being, as Bayard de Volo argues, the use of weaponised drones erodes traditional military masculinity and the identity of the warrior. The result, some commentators argue, is a 'feminisation', associated with the degradation of the using militaries (Coker, 2002; Van Creveld, 2013). Reflecting the construction of masculinity and femininity as mutually oppositional, the other side of the argument is that the use of drones produce precisely the opposite effect – that is the 'hyper-masculinisation' – of the using militaries (Kunashakaran, 2016).

To illustrate how gender, and specifically the distinction between masculinity and femininity, functions in relation to drones it is necessary to take a brief detour to discuss in more detail the importance of this binary[10] construction to thinking about gendering military technologies. The co-constitution of gender and war can be explored through the designation of masculinity and femininity as a hierarchical binary so that the terms can only be understood in relation to each other and as mutually exclusive. For example, masculinity is what is not femininity and vice versa (Shepherd, 2010; Steans, 2013). Also, because the binary is hierarchical, masculinity is valorised, meaning that those things associated with femininity are subordinated to those associated with masculinity (Carver, 2006). Peterson and Runyan describe this as the problem of 'the status of dichotomies' whereby the masculine attributes are 'considered desirable, admirable, and preferred whereas … [the feminine] are variously understood as less desirable, inferior, and/or threatening to the esteemed qualities of [the masculine]' (Peterson and Runyan, 2010, p. 46).

Dichotomies (or binaries) are powerful because they provide short cuts for thinking, and this means that they are used often, and the assumptions that they are founded on become taken for granted and naturalised. Because binaries offer us only two choices, and two choices that are mutually exclusive, they 'forestall our consideration of non-oppositional constructions [such as] right in relation to plausible, persuasive, preferable, viable; rational in relation to consistent, instrumental' etc., closing down a whole host of possible alternatives because we become locked into the pattern of thinking either/or (Peterson and Runyan, 2010, p. 50). Similarly, because we see the two choices presented as polar opposites, we forget to see the hierarchy that is implicit between them and their contingency on a specific cultural and historical setting so that they become construed as 'natural': timeless and pre-existing social interaction. This apparent fixedness makes it difficult to question binaries as claims to naturalness imply that change is neither possible nor desirable. Poststructural feminists have sought to acknow-ledge the political power of binary labels whilst simultaneously questioning them.

Jean Bethke Elshtain's (1995) work on the 'Just Warrior' and 'Beautiful Soul' illustrates how the ideas of masculinity and femininity are mapped onto institutions like the military, and how the legitimacy bestowed on the military in its enacting of violence reflects the ideas of what is and is not gender 'appro-priate' (Sjoberg, 2010a). The military body, rather than military bodies, is con-structed as masculine; it is 'contained' and impenetrable, 'ready for action, and always under control' (Taylor, 1993, p. 22). Not for the military the 'weak and leaky' feminine body (Basham, 2013). According to traditional narratives, women become the reason for going to war, but are not considered appropriate figures to conduct the war. Women can be victims of war, but never perpetrators. As a result, women's violence is pathologised where men's is, within the frame of war, celebrated as heroic (Elshtain, 1995; Sjoberg, 2007; Goldstein, 2001). This is not to argue that women are less violent and that men are more violent, rather that in discourses of war masculinity is associated with action, strength, and violence, and femininity with passivity, victimhood, and weakness (Braudy, 2005, p. 328).

One way of exploring the power of these binaries in discourses in war, and their limitations, is through the disruptive figure of the female solider. Because traditional security narratives essentialise the connection between sex and gender (that is, they implicitly connect being a man with masculinity and being a woman with femininity) discourses surrounding women soldiers are complicated and confused (Howard III and Prividera, 2004; Just, 2006; MacKenzie, 2012).

Feminist scholars have illustrated this confusion through the narratives used to tell the story of the experiences of Private Jessica Lynch. Lynch was captured by Iraqi soldiers in 2003 and was 'rescued' by (all male) US forces after nine days of captivity (Just, 2006). Lynch's capture and subsequent rescue provided two key moments that reinforced traditional gender discourses. First, as a (female) prisoner of war, in the hands of dangerous (sexually deviant) 'brown' men, commentators speculated that she was at risk of being raped – (re)inscrib-ing her body as both sexual and open to penetration (Richter-Montpetit, 2007). The presence of 'brown' captors served to heighten concerns about sexual assault, and so powerful was this narrative that even after Lynch subsequently vociferously denied any sexual harassment by her captors, the story continued to circulate (Pin-Fat and Stern, 2005). Second, by 'getting herself captured', Lynch established herself in the (feminine) role of 'passive victim' and enabled Amer-ican soldiers to reaffirm their masculinity through enacting the role of her heroic protectors and rescuers (Howard III and Prividera, 2004; Pin-Fat and Stern, 2005). From the accounts of Jessica Lynch as both a soldier and a 'victim', it becomes clear that her position as both masculine (soldier) and feminine (victim) in dominant discourses is difficult to navigate.

I am interested in the way that Lynch's story was told because the discomfort with the erosion of gendered boundaries is also at work in the lives of British

Reaper crews. Additionally, whilst Lynch's experiences can and have been viewed as reinforcing patriarchal hierarchies in the US military, British Reaper crews can be viewed as *both* reinforcing *and* destabilising traditional narratives of military masculinity. Therefore, having outlined the importance of acknowledging the role that dichotomous thinking plays in discourses of gender and sketched out some of the ways that feminist theorising uses this to illuminate topics relating to warfare and military technologies, I turn now to feminist interjections that deal specifically with the military technology of 'drones'. I begin by outlining the way in which the debate is often framed in terms of whether the use of drones acts as a feminising or masculinising influence on the using militaries. I then argue that this thinking is limited and cannot explain the ways in which drones are *simultaneously both* feminising and masculinising. In order to do so I draw on important queer scholarship on drone warfare in order to suggest an alternative methodological framework.

Gendered logics: feminisation/masculinisation

In order to outline the arguments that drones erode masculinity and result in the 'feminisation' of the militaries that use drones it is necessary to quickly layout what I mean when I use the term 'feminisation'. Whilst it is often connected to the introduction of women to the military sphere (beyond the traditional roles of camp follower, adoring stay-at-home wife etc.), 'feminisation' is also used to describe the degradation of the military (the work it does, the people who are part of it etc.) (Coker, 2000). Given that war is the traditional preserve of the male (as well as the masculine) the introduction of women into this hallowed space is associated with the introduction of the lesser-valued/degrading feminine (Mitchell, 1989; Van Creveld, 2001, 2013). As Kimberly Hutchings aptly phrases it,

> feminization has a double meaning. On one hand, feminization refers to the fact that there are not many women in the professional militaries of advanced industrial societies; on the other hand, feminization refers to a process of decline in the capacity to engage in so-called real war.
>
> (2007, p. 395)

Feminisation, then, may be understood as a process of moving towards a constellation of attributes that compare unfavourably with the masculine. In relation to war this includes attributes that are not considered conducive to being able to successfully wage war – cowardice, weakness, emotionality. In relation to technologies of war, those technologies that might be considered 'feminising' are those that encourage (or at least do not discourage) the feminine attributes listed above.

The opposite to feminisation is masculinisation which may be understood as the process of increasing in traits associated with masculinity and it is worth pausing here for a moment to take a brief detour through that concept. Paul

Higate summarises masculinity as 'a diverse cluster of relational values, beliefs, performances and ideologies' (2012, p. 31), and as such, is 'not a coherent object' (Connell, 2005, p. 67). However, given its relationship to femininity, we can understand masculinity as those things which femininity is not: bravery, strength, rationality (Higate, 2012, p. 31; Hutchings, 2007). This is not to argue, as I have noted before, that *women* cannot be brave, strong or rational or that *men* cannot be cowardly, weak or emotional, but rather than these traits have become part of the stories that we tell about what it *means* to be a man or a woman (Sylvester, 1994).

There are many different kinds of masculinity[11] and recent work in this area has revealed a complex and shifting web of *masculinities* (for example, Connell, 2005; Hooper, 2001; Brittan, 1989; Barrett, 1996; Zalewski and Parpart, 1998). This book is concerned with military masculinity/ies and how masculinity is implicated in the experiences of war (Hockey, 2003; Higate, 2012; Barrett, 1996). This/These masculinity/ies reflect the particular experiences and landscape of the military and its interaction with gender (Goldstein, 2001), and refers to 'a wide range of behaviours including aggressive, rational, courageous, cool, calculating, chivalrous, [and] protective' (Higate, 2012, p. 32; see also Morgan, 1994, pp. 391–392). Masculinisation in war is then the process through which these particular traits become valorised and more nearly associated with the appropriate military performance of masculinity, as Coker notes 'warriors are no good if they are emasculated' (2013, p. 48).

I refer in this book to masculinity, particularly in relation to military masculinity, in the singular not because I perceive it to be monolithic, but rather to highlight the relationship between masculinity and femininity which forms the primary dyad around which this book is constructed. Additionally, I refer to military masculinity over masculinities because I am referring to the particular military masculinity performed by British Reaper crews. Finally, I utilise masculinity in the singular simply to maintain grammatical and textual simplicity, to avoid continually references to 'This/These masculinity/ies' as written above. In reading the singular 'masculinity' (and of course femininity) throughout the rest of this book, I ask you as the reader to keep in mind the multidimensional nature of the concept and the implications of this, which is extensively explored and reviewed in the scholarship into/on masculinities (for example Connell, 2005; Hooper, 1998; Donaldson, 1993; Whitworth, 2005; Lee, 2009; McCarry, 2007).

From this definitional detour, let us return to the topic at hand: the gendered logics applied to the military use of drones. Commentators who decry the use of drones because they see them as eroding military masculinity are not always explicit about the connection with gender. Concern about the lack of (physical) risk to drone crews is connected to a concern about an attendant loss of valour, courage, and heroism as is explored in more detail in Chapter 4 (Coker, 2013; Vallor, 2013; Baggiarini, 2015). As the previous sections have illustrated, these are all attributes associated with masculinity, and particularly with military masculinity. The 'armchair killers' of Colin Freeman's (2015) report (also referred to as 'the chair force' (Gusterson, 2014; Power, 2013)) are uncomfortable figures because they do not conform to our idea of what it means to be a soldier or a warrior. Rather, they spend all day sat in (arm)chairs watching video feeds called 'predator porn' (Freeman, 2015). Similarly, the physical environment in which drone crews operate, Manjikian writes, is 'pure and unsullied', a description 'which has historically been coded as feminine' and which is the polar opposite to the usual chaos, dirt, and squalor associated with war, which we will return to in Chapter 6 (2013, p. 52; see also Peterson and Runyan, 2010; Kunashakaran, 2016). Wilcox takes this further by arguing that

> The drone operator is also 'unmanned' because he is frequently stationed not in the war zone, but in the feminized space of the 'home front' and is kept out of danger.
>
> (2015, p. 135)

Whilst I will challenge the perception of risklessness in Chapter 4, this narrative is an important one to arguments of feminisation and one which I elaborate in the following paragraphs.

There are concerns that killing by drone is too easy (Coeckelbergh, 2013; Sauer and Schornig, 2012). The ease of killing is both a moral question (is it unethical for killing to be easier? (Grossman, 2009; Cole *et al.*, 2010)) and a gendered one – easy killing does not require courage or physical strength, it erodes the status of those individuals who have risked life and limb in order to undertake what is considered to be a difficult and hallowed task (Kaurin, 2014; Coker, 2002). The drone crews, it is argued, erode the status of the warrior by killing at the push of a button; they do not have to get down and dirty, grapple with the enemy or look him (or her) squarely in the eye before killing him (Mayer, 2009; Asaro, 2013; Royakkers and van Est, 2010). For example, infantry veteran Scott Beauchamp (2016) argues that the lack of 'existential tension' in the lives of Reaper crews renders this format of war a 'particular form of martial law' rather than *real* conflict. This kind of argument, that drone warfare is something less than, or other than *real* warfare (and therefore the individuals who conduct it are not *real* warriors) is reflected by a number of other authors, particularly those who connect warfare with physical risk (see Chapter 4 for discussion) (Chamayou and Lloyd, 2015; Luttwak, 1995; Riza, 2013; Mayer, 2009)

The symbolic importance of the warrior's body is central to concerns that using drones will result in the feminisation of the armed forces that use them. Without risking his body the warrior does not have to overcome his fear of injury and death, which it is argued is central to the position of war as 'the traditional test of manhood' (see Chapter 3 for detailed discussion on the warrior figure) (Coker, 2013, p. 73). The test is passed, Coker argues, by 'standing firm under fire, man to man', supporting your comrades by not running away (or being a coward) and presumably by not dying (2013, p. 73; see also Duncanson, 2007; Morgan, 1994; Hockey, 2003). Risk is a 'a key component of both hegemonic masculinity (Hinojosa 2010) and the warrior's identity', and the (apparent) lack of physical risk to drone crews is a barrier to their performance of masculinity (Manjikian, 2013, p. 52). I explore this issue of risk and lack of risk in relation to military masculinity and the warrior identity in Chapter 3 but suffice to say here that the inability of drone crews to risk life and limb is a core component of the arguments of those who are concerned about feminisation.

Counter to the claims of those who are concerned about feminisation are those scholars who see the drone as threatening because it heralds, or highlights, hyper-masculinisation in the armed forces. Sumita Kunashakaran (2016), who makes this argument, views hyper-masculinisation as a process through which those traits associated with masculinity are *particularly* valorised, to an extent that is even greater than usual. In Kunashakaran's (2016) piece she argues for this hyper-masculinisation on the basis of two elements: first the identification of the drones themselves as hyper-masculine and second, on the basis of the crews becoming hyper-masculinised.

The drones themselves are construed as hyper-masculine because of their disconnection from the physical bodies of their crews in two main ways. First, with their lack of an on-board pilot, drones have 'the ability to operate in the field for lengthened periods of time without expensive redundancies and life-support systems' (Kunashakaran, 2016, p. 44). Without a body on board, without the need to cater to the requirements of the flesh, and carrying complex data gathering and processing equipment, the drone becomes the equivalent of a flying mind, the embodiment of the masculine rational. Indeed, Heather Roff claims that the 'robot fighter is the ideal of masculinity in western culture, for it represents an "independent, risk-taking, aggressive, heterosexual and rational" being free from any weakness …' (2016, p. 2). Therefore, where Haraway (1991) argues for the cyborg as post-human and post-gender, others instead view the 'cyborg solider' as the point at which '*technology* embodies *masculinity*', rendering the human soldier the feminised other (Masters, 2005, p. 113).

Second, hypermasculinity is also constructed through the drone's capacity to 'stare'. As is explored in more detail in Chapter 5 persistent surveillance by the drone (and its human crews) situates the drone as the masculine watcher in relation to the feminised spectacular 'Other' on the ground (Holmqvist, 2013; Kindervater, 2016; Wall and Monahan, 2011; Carver, 2008; Smith, 2016). Additionally, the 'vision' of the drone is beyond human capabilities – with the ability to see in the dark and via infra-red, the technologies outstripping the efforts of even 'the best' warriors. As such the drone demonstrates the masculine traits of 'efficiency and proficiency' in ways that the human warriors are not able to do so (Kunashakaran, 2016, p. 54).

Whilst it is interesting to investigate the drone *machinery* as masculine, this book is more concerned with Kunashakaran's (2016) second argument, which centres around the hyper-masculinisation of the crews themselves. Here the argument is that the physical remoteness of the drone crews results in emotional remoteness, analogous to the stoicism and emotional reserve construed as a hyper-masculine warrior trait (Higate, 2012; Coker, 2007). Masculinity is also conveyed through the position of the drone *above* the individuals who are being watched, and as such this can be interpreted as not just confirming the masculinity of the drone, but also of the crews who are privileged by this 'god's eye view' (see Chapter 5 for discussion). The 'drone stare' and the 'subjugation of those marked Other', reaffirms the masculine status of the crews who are situated in the position of power watching those 'Others' who become the feminised spectacular other (Holmqvist, 2013, p. 452). Similarly, as Wilcox argues,

> the 'disembodiment' of the pilot also means that he is not confined to the particularities and limited vision of his body; the satellite systems and the drone's video cameras mean that the bomber's eye view is the God's eye view of objectivity.
>
> (Wilcox, 2015, p. 135)

Here Wilcox disrupts Kunashakaran's distinction between the two masculinising components, arguing instead that they are two sides of the same coin. In this understanding the drone crews' masculinity is established on the basis of two elements – overcoming the limits of the physical body (which is traditionally coded as feminine) and in being able to act with 'objectivity' (aligned with the masculine side of the rational/emotional binary).

The traits that are considered desirable in drone crews, 'situational awareness and strategic thinking skills' are, Kunashakaran argues, 'both highly hegemonically masculinized traits' (2016, p. 44; see also Kontour, 2012; McKinley *et al.*, 2011). Technological know-how has not always necessarily been connected with hyper-masculinity (see for example Coker, 2013, p. 116) but, as militaries (particularly Western militaries) have embraced the potential benefits of novel technologies, technological proficiency has become part of the masculinity constellation, demonstrating the social construction and fluidity of masculinity (Manjikian, 2013; Kontour, 2012; Barrett, 1996). As Braudy notes, 'Technology would not undermine either war or masculinity but enhance and perfect them … the best men would be matched with the best machines, "for the two were inseparable"' (2005, p. 403)

Finally, the *way* that drones are used and the types of missions that they are used to undertake may also be understood as masculine. Drones have been used to launch lethal strikes as part of the 'Global War on Terror', and feminist scholars have highlighted the many ways in which this conflict has been constructed as an intervention by chivalrous white men to save brown women from dangerous brown men, re-emphasising the continuing salience of the protector/ protected dyad (to be clear, this is not to argue that this is the *reality* but rather that these are the discourses that circulate in debates about this conflict) (for example Shepherd, 2006, 2008b; Richter-Montpetit, 2007; Sjoberg, 2010a; Bhattacharyya, 2008; Masters, 2009). Similarly, and perhaps *because* of these narratives, the ability of drones to penetrate deep into the sovereign territories of 'brown men', brings narratives of sexuality to the fore. References to sodomisation emerge in narratives which reinforce ideas of dominance and submission, and reflect the feminisation of the submissive individual(s)/states (Richter-Montpetit, 2007; Puar, 2005; Hooper, 2001).

In the narratives about masculinisation and feminisation through the use of drones, many of the accounts acknowledge data which does not, or cannot be made, to fit with these *either/or* positions. To clarify, the accounts concerned with feminisation often note that there are elements of masculinisation and the accounts that argue the use of drones will result in masculinisation include references to feminisation. For example, Kunashakaran's account of the masculinisation includes references to ' "feminized" skills', the lack of risk and the way the 'UAV control rooms are a far cry from the grimy battlefields' (2016, p. 40). Similarly, the concerns about feminisation are often couched in terms lack of physical risk, but are unsettled by references to emotional or psychological distance, traditionally construed as masculine (Wilcox, 2015; Holmqvist, 2013). Whilst it is possible that these instances act as the exceptions that prove the rule, the anomalous elements are important components of the lives and experiences of the Reaper crews which cannot be easily brushed under the carpet. That the 'feminized skills' sit (un/comfortably?) alongside narratives of the masculinity of technological mastery; or that the distance creates emotional reserve and ease of killing, whilst *at the same time* persistent surveillance creates a strange sense of intimacy which is traumatising to crews – indicates that there is something missing or being missed in the existing frameworks that are applied to this topic of research. There is a space in the literature for a framework and methodology that can take all of these confusing, contradictory pieces and make better sense of them. To do this, the next chapter will introduce the methodological framework of Haunting.

Conclusion

This chapter summarised the way in which military scholars have traditionally addressed the interaction between warfare and technology. Tracing the lines between scholars who have tended to ignore the implications of military developments and those who have bordered on embracing technological determinism, I have illustrated the shapes of the debates in these areas. However, I have argued that even in those accounts that acknowledge the importance of the interaction between the technology and the humans designing, creating and wielding it, traditional security scholars have neglected to consider the way that individual differences affect this interaction. Feminist security scholars offer an important interjection to this debate by demonstrating the myriad ways in which gender and war, and therefore gender and the technologies of war are inter-related and co-constituting. Feminist security scholarship does not only indicate the importance of putting women back into the picture drawn of global politics and military affairs, but also the need to gender men by acknowledging the way discourses of war are gendered.

Drawing attention to feminist scholarship at the particular nexus of war and technology, I have illustrated the ways in which, from the specific weapons to the way that they are used and the way(s) that that use is spoken of, this nodal point is woven through with gendered discourses, symbolism, and meaning.

Turning to the issue of the use of robots in war, feminists have argued for both their emancipatory potential and their capacity to re-inscribe traditional masculine narratives. These debates formed the basis from which moved to consider the power of the oppositional binary and the way in which this logic is implicated the work that gender does in relation to war and military technologies. Using gendered binaries limits the ways we think and shut down opportunities for alternative modes of thinking. In the debate surrounding the use of drones, masculinity and femininity were (sometimes implicitly and sometimes explicitly) mobilised in arguments which either support or critique the use of this technology. Importantly, in sketching out the masculinisation *or* feminisation arguments I have drawn attention to the ways in which this dichotomy is incomplete and the need for a framework and methodology that can make sense of the data that does not fit in with the current ways of thinking. The framework that I utilise to fill this 'gap' in thinking, and which forms the primary contribution of this book, is that of 'Haunting' which I outline in detail in the following chapter.

Notes

1 (Although this terminology is contested, see Sloan, 2002 for debate; Hundley, 1999; Neuneck, 2008; Rogers, 1995).
2 The issue of autonomous weapons is one that for now remains theoretical, and is beyond the primary concerns of this book, embedded in its own huge, detailed and hotly debated literature(s).
3 In keeping with academic conventions, I capitalise 'International Relations' when I refer to the discipline but use lower case for other references.
4 This has also been a critique of work within the sphere of feminist security studies, whereby academics have appeared to try to speak for all women or for all feminists (Marchand, 1998; Steans, 2013; Peterson, 1992; Pettman, 1996; Mohanty, 2003).
5 For example see (Keohane, 1989; Jones, 1996).
6 I want to avoid implying that 'IR has been and continues to be wholly dominated by positivists' who brook no counter argument, however, following Jill Steans, it is 'meaningful to speak of the dominance of realism and latterly neo-realism within the field ...' (2003, p. 429).
7 Indeed conceptualising 'gender' as synonymous with women ignores a vast and vibrant body on masculinities (for example, Connell, 2005; Hooper, 2001; Hockey, 2003; Allsep, 2013; Barrett, 1996; Baaz and Stern, 2009; Ashworth and Swatuk, 1998; Zalewski and Parpart, 1998).
8 Just as 'men are not synonymous with humanity but have a socially constructed gender with no special claim to physical or mental superiority...' (Ramazanoglu, 1992, p. 340).
9 Cohn recalls one male physicist blurting out in a group meeting 'Wait, I've just heard how we're talking – *Only* 30 million! *Only* 30 million human beings killed instantly?', he recalled later that 'Silence fell upon the room. Nobody said a word. They didn't even look at me. It was awful. I felt like a woman' (Cohn *et al.*, 2006, p. 4).
10 The term binary refers to two categories that are understood to be mutually exclusive. I tend to use the term binary or binaries, but occasionally use, after other authors, dyad or dichotomy. I use these terms interchangeably and to imply the same thing.
11 This is also true of femininity, although this is not the focus of the book.

References

Afshar, H. and Eade, D. (eds) (2004) *Development, women, and war: feminist perspectives*. Development in practice reader. Oxford: Oxfam.

Allsep, L.M. (2013) The myth of the warrior: martial masculinity and the end of don't ask, don't tell. *Journal of Homosexuality*, 60 (2–3): 381–400. doi: 10.1080/00918369.2013.744928.

Anderson, K. (2000) Ottawa convention banning landmines, the role of international non-governmental organizations and the idea of international civil society. *European Journal of International Law*, 11: 91.

Asaro, P.M. (2008) How just could a robot war be? In *Proceedings of the 2008 Conference on Current Issues in Computing and Philosophy*. Peace Research Institute, Oslo, 2008. Security Dialogue. pp. 50–64.

Asaro, P.M. (2013) The labor of surveillance and bureaucratized killing: new subjectivities of military drone operators. *Social Semiotics*, 23 (2): 196–224. doi: 10.1080/10350330.2013.777591.

Ashworth, L. and Swatuk, L. (1998) Masculinity and the fear of emasculation in international relations theory. In Zalewsk, Marysia (ed.) *The 'man question' in international relations*. Builder, CO: Westview Press.

Baaz, M.E. and Stern, M. (2009) Why do soldiers rape? Masculinity, violence, and sexuality in the armed forces in the Congo (DRC). *International Studies Quarterly*, 53 (2): 495–518. doi: 10.1111/j.1468-2478.2009.00543.x.

Baaz, M.E. and Stern, M. (2011) Whores, men, and other misfits: undoing 'feminization' in the armed forces in the DRC. *African Affairs*, 110 (441): 563–585. doi: 10.1093/afraf/adr044.

Baggiarini, B. (2015) Drone warfare and the limits of sacrifice. *Journal of International Political Theory*, 11 (1): 128–144. doi: 10.1177/1755088214555597.

Barkawi, T. (2011) From war to security: security studies, the wider agenda and the fate of the study of war. *Millennium – Journal of International Studies*, 39 (3): 701–716. doi: 10.1177/0305829811400656.

Barnaby, F. (ed.) (1984) *Future war: armed conflict in the next decade*. London: Joseph.

Barrett, F.J. (1996) The organizational construction of hegemonic masculinity: the case of the US Navy. *Gender, Work & Organization*, 3 (3): 129–142. doi: 10.1111/j.1468-0432.1996.tb00054.x.

Basham, V. (2013) War, *identity and the liberal state: everyday experiences of the geopolitical in the Armed Forces. Interventions*. 1st edn. London; New York: Routledge.

Bayard de Volo, L. (2013) Unmanned? Drones and the revolution in gender-military affairs. *Conference Paper for 3rd European Conference on Politics and Gender*. Available at: www.ecpg-barcelona.com/sites/default/files/Ppr-Unmanned-ECPG.pdf.

Bayard de Volo, Lorraine (2016) Unmanned? Gender recalibrations and the rise of drone warfare. *Politics & Gender*, 12 (1): 50–77.

Baylis, J. and O'Neill, R.J. (eds) (2000) *Alternative nuclear futures: the role of nuclear weapons in the post-cold war world*. Oxford; New York: Oxford University Press.

Beason, D. (2005) *The E-bomb: how America's new directed energy weapons will change the way future wars will be fought*. 1st Da Capo Press edn. Cambridge, MA: Da Capo Press.

Beauchamp, S. (2016) Can drone pilots be heroes? *The Atlantic*, 23 January. Available at: www.theatlantic.com/politics/archive/2016/01/can-drone-pilots-be-heroes/424830/ (accessed: 27 January 2016).

Bhattacharyya, G. (2008) *Dangerous brown men: exploiting sex, violence and feminism in the war on terror*. London; New York: Zed Books; Distributed in the USA exclusively by Palgrave Macmillan.

Boot, M. (2007) *War made new: weapons, warriors, and the making of the modern world*. A Council on Foreign Relations book. First trade paperback printing, August 2007. New York: Gotham Books.

Braudy, L. (2005) *From chivalry to terrorism: war and the changing nature of masculinity*. 1. Vintage Books edn. New York: Vintage Books.

Brigety, R.E. (2007) *Ethics, technology, and the American way of war: cruise missiles and US security policy*. Contemporary security studies. London; New York: Routledge.

Brittan, A. (1989) *Masculinity and power*. Oxford, UK; New York, USA: Basil Blackwell.

Burgess, J.P. (ed.) (2010) *The Routledge handbook of new security studies*. Routledge handbooks. London; New York: Routledge.

Butler, J. (1999) *Gender trouble feminism and the subversion of identity*. New York: Routledge. Available at: http://kcl.etailer.dpsl.net/home/html/moreinfo.asp?isbn=0203 902750&whichpage=1&pagename=category.asp (accessed: 17 September 2014).

Butler, J. (2011) *Bodies that matter: on the discursive limits of 'sex'*. Routledge classics. Abingdon, Oxon; New York: Routledge.

Carver, T. (1996) *Gender is not a synonym for women*. Gender and political theory. Boulder, CO: L. Rienner.

Carver, T. (2006) Being a man. *Government and Opposition*, 41 (3): 450–468. doi: 10.1111/j.1477-7053.2006.00187.x.

Carver, T. (2008) Men in the feminist gaze: what does this mean in IR? *Millennium – Journal of International Studies*, 37 (1): 107–122. doi: 10.1177/0305829808093767.

Casey-Maslen, S. (ed.) (2014) *Weapons under international human rights law*. Cambridge, UK; New York: Cambridge University Press.

Chamayou, G. and Lloyd, J. (2015) *Drone theory*. New York: Penguin.

Clark, I. (2015) *Waging war: a new philosophical introduction*. Oxford: Oxford University Press. Available at: www.oxfordscholarship.com/view/10.1093/acprof:oso/978019 8724650.001.0001/acprof-9780198724650 (accessed: 12 September 2016).

Coeckelbergh, M. (2013) Drones, information technology, and distance: mapping the moral epistemology of remote fighting. *Ethics and Information Technology*, 15 (2): 87–98. doi: 10.1007/s10676-013-9313-6.

Cohn, C. (1987) Sex and death in the rational world of defense intellectuals. *Signs*, 12 (4): 687–718.

Cohn, C., Hill, F. and Ruddick, S. (2006) *The relevance of gender for eliminating weapons of mass destruction*. 38. Stockholm, Sweden: Weapons of Mass Destruction Commission. Available at: www.un.org/disarmament/education/wmdcommission/files/No38.pdf (accessed: 19 September 2014).

Coker, C. (2000) Humanising warfare, or why Van Creveld may be missing the 'big picture'. *Millennium – Journal of International Studies*, 29 (2): 449–460. doi: 10.1177/03058298000290020201.

Coker, C. (2002) *Waging war without warriors?: the changing culture of military conflict*. IISS studies in international security. Boulder, CO: Lynne Rienner Publishers.

Coker, C. (2004) *The future of war: the re-enchantment of war in the twenty-first century*. Blackwell manifestos. Malden, MA; Oxford, UK: Blackwell Pub.

Coker, C. (2007) *The warrior ethos: military culture and the war on terror*. LSE international studies series. London; New York: Routledge.

Coker, C. (2013) *Warrior Geeks*. London: C Hurst & Co Publishers Ltd.

Cole, C., Dobbing, M. and Hailwood, A. (2010) *Convenient killing: armed drones and the 'Playstation' mentality*. Available at: http://dronewarsuk.files.wordpress.com/2010/10/conv-killing-final.pdf (accessed: 24 April 2013).

Connell, R.W. (2005) *Masculinities*. 2nd edn. Berkeley, CA: University of California Press.

Cook, Tim (2000) 'Against God-inspired conscience': the perception of gas warfare as a weapon of mass destruction, 1915–1939. *War and Society*, 18 (1): 47–69.

Daggett, C. (2015) Drone disorientations: how 'unmanned' weapons queer the experience of killing in war. *International Feminist Journal of Politics*, 17 (3): 361–379.

Danielson, P. (2011) Engaging the public in the ethics of robots for war and peace. *Philosophy & Technology*, 24 (3): 239–249. doi: 10.1007/s13347-011-0025-8.

D'Costa, B.D. (2006) Marginalized identity: new frontiers of research for IR? In Ackerly, B.A. and Stern, M. (eds) *Feminist Methodologies for International Relations*. Cambridge: Cambridge University Press. pp. 129–152.

Donaldson, M. (1993) What is hegemonic masculinity? *Theory and Society*, 22 (5): 643–756.

Doty, R.L. (1996) *Imperial encounters: the politics of representation in North-South relations*. Borderlines v. 5. Minneapolis, MN: University of Minnesota Press.

Duncanson, C. (2007) *Forces for good? British military masculinities on peace support operations*. Edinburgh: AIAA.

Echevarria, A.J. (2007) *Imagining future war: the West's technological revolution and visions of wars to come, 1880–1914*. War, technology, and history. Westport, CT: Praeger Security International.

Elshtain, J.B. (1995) *Women and war*. University of Chicago Press ed. Chicago, IL: University of Chicago Press.

Engels, F. (1976) *Engels as military critic: articles*. Westport, CT: Greenwood Press.

Enloe, C.H. (1993) *The morning after: sexual politics at the end of the Cold War*. Berkeley, CA: University of California Press.

Enloe, C.H. (2000) *Bananas, beaches and bases: making feminist sense of international politics*. Berkeley, CA: University of California Press.

Enloe, C.H. (2013) *Seriously! Investigating crashes and crises as if women mattered*. Berkeley, CA: University of California Press.

Evangelista, M., Shue, H., and Biddle, T.D. (eds) (2014) *The American way of bombing: changing ethical and legal norms, from flying fortresses to drones*. Ithaca, NY; London: Cornell University Press.

Freeman, C. (2015) Armchair killers: life as a drone pilot. *Telegraph*, 4 November. Available at: www.telegraph.co.uk/culture/film/film-news/11525499/good-kill-drone-pilot-true-story.html (accessed: 30 June 2015).

Futter, A. (2011) NATO, ballistic missile defense and the future of US tactical nuclear weapons in Europe. *European Security*, 20 (4): 547–562. doi: 10.1080/09662839.2011.626404.

Galamas, F. (2008) Biological weapons, nuclear weapons and deterrence: the biotechnology revolution. *Comparative Strategy*, 27 (4): 315.

Gartzke, E. (2013) The myth of cyberwar: bringing war in cyberspace back down to earth. *International Security*, 38 (2): 41–73.

Geiβ, R. (2011) Poison, gas and expanding bullets: the extension of the list of prohibited weapons at the review conference of the International Criminal Court in Kampala. In Schmitt, M.N., Arimatsu, L., and McCormack, T. (eds) *Yearbook of international humanitarian law – 2010*. The Hague, The Netherlands: T.M.C. Asser Press. pp. 337–352.

Available at: www.springerlink.com/index/10.1007/978-90-6704-811-8_11 (accessed: 19 May 2015).

Goldstein, J.S. (2001) *War and gender: how gender shapes the war system and vice versa*. Cambridge: Cambridge University Press.

Gray, C.D. (2008) *International law and the use of force*. Foundations of public international law. 3rd edn. Oxford; New York: Oxford University Press.

Gray, C.H. (ed.) (1995) *The cyborg handbook*. New York: Routledge.

Gronnvoll, M. and McCauliff, K. (2013) Bodies that shatter: a rhetoric of exteriors, the abject, and female suicide bombers in the 'War on Terrorism'. *Rhetoric Society Quarterly*, 43 (4): 335–354. doi: 10.1080/02773945.2013.819989.

Grossman, D. (2009) *On killing: the psychological cost of learning to kill in war and society*. Rev. ed. New York: Little, Brown and Co.

Gusterson, H. (2014) Toward an anthropology of drones: remaking space, time and valor in combat. In Evangelista, M. and Shue, H. (eds) *An American way bombing: changing ethical and legal norms from B-17s to drones*. Ithaca, NY; London: Cornell University Press. pp. 191–206.

Hamilton, S.N. (1997) The cyborg, 11 years later: the not-so-surprising half-life of the cyborg manifesto. *Convergence: The International Journal of Research into New Media Technologies*, 3 (2): 104–120. doi: 10.1177/135485659700300211.

Hansen, L. (2009) *Evolution of International Security Studies*. Cambridge: Cambridge University Press. Available at: www.myilibrary.com?id=239375 (accessed: 5 October 2014).

Haraway, D. (1991) A cyborg manifesto: science, technology and socialist feminism in the late twentieth century. In *Simians, Cyborgs and Women: The Reinvention of Nature*. New York: Routledge. pp. 149–181.

Higate, P. (2012) Foregrounding the in/visibility of military and militarised masculinities. In Eriksson Baaz, Maria and Utas, Mats (eds) *Beyond gender and stir: reflections on gender and SSR in the aftermath of African conflicts*. Uppsala: The Nordic Africa Institute. pp. 31–37.

Hinojosa, Ramon (2010) Doing hegemony: military, men, and constructing a hegemonic masculinity. *The Journal of Men's Studies*, 18 (2): 179–194.

Hockey, J. (2003) No more heroes: masculinity in the military. In Higate, P. (ed.) *Military masculinities*. Westport CT: Greenwood Publishing Group.

Holmqvist, C. (2013) Undoing war: war ontologies and the materiality of drone warfare. *Millennium – Journal of International Studies*, 41 (3): 535–552. doi: 10.1177/0305829813483350.

Hooper, C. (1998) Multiple masculinities in international relations. In Zalewski, M. (ed.) *The 'man question' in international relations*. Boulder, CO: Westview Press.

Hooper, C. (2001) *Manly states: masculinities, international relations, and gender politics*. New York: Columbia University Press.

Howard III, John W. and Prividera, Laura C. (2004) Rescuing patriarchy or saving 'Jessica Lynch': the rhetorical construction of the American woman soldier. *Women and Language*, 27 (2): 89–97.

Hundley, R.O. (1999) *Past revolutions, future transformations: what can the history of revolutions in military affairs tell us about transforming the U.S. military?* Santa Monica, CA: Rand.

Hutchings, K. (2007) Making sense of masculinity and war. *Men and Masculinities*, 10 (4): 389–404. doi: 10.1177/1097184X07306740.

IKV Pax Christi (2011) *Does unmanned make unacceptable? Exploring the debate on using drones and robots in warfare.* Utrecht, The Netherlands: IKV Pax Christi.

Immerman, R.H. and Goedde, P. (eds) (2013) *The Oxford handbook of the Cold War.* Oxford handbooks. 1st edn. Oxford, UK: Oxford University Press.

Jensen, G. (2001) *The meaning of war in a technological age.* In Jensen, G. and Wiest, A. (eds) New York: New York University Press.

Johnson, A.M. and Axinn, S. (2013) The morality of autonomous robots. *Journal of Military Ethics*, 12 (2): 129–141. doi: 10.1080/15027570.2013.818399.

Jomini, A.H. (2005) *The art of war.* El Paso, TX: El Paso Norte Press.

Jones, A. (1996) Does 'gender' make the world go round? Feminist critiques of international relations. *Review of International Studies*, 22 (04): 405. doi: 10.1017/S0260210500118649.

Just, S.N. (2006) Embattled agencies – how mass mediated comparisons of Lynndie England and Jessica Lynch affect the identity positions available to female soldiers in the US army. *Scandinavian Journal of Management*, 22 (2): 99–119. doi: 10.1016/j.scaman.2006.03.006.

Kaag, John and Kaufman, Whitley (2009) Military frameworks: technological know-how and the legitimization of warfare. *Cambridge Review of International Affairs*, 22 (4): 585–606.

Kaurin, P. (2010) With fear and trembling: an ethical framework for non-lethal weapons. *Journal of Military Ethics*, 9 (1): 100–114. doi: 10.1080/15027570903523057.

Kaurin, P.M. (2014) *The warrior, military ethics and contemporary warfare: Achilles goes asymmetrical.* Military and defence ethics. Farnham, Surrey: Ashgate Publishing Company.

Kennedy-Pipe, C. (2000) Women and the military. *Journal of Strategic Studies*, 23 (4): 32–50. doi: 10.1080/01402390008437811.

Keohane, R.O. (1989) International relations theory: contributions of a feminist standpoint. *Millennium – Journal of International Studies*, 18 (2): 245–253. doi: 10.1177/03058298890180021001.

Kindervater, K.H. (2016) The emergence of lethal surveillance: watching and killing in the history of drone technology. *Security Dialogue.* doi: 10.1177/0967010615616011.

Kontour, K. (2012) The governmentality of battlefield space: efficiency, proficiency, and masculine performativity. *Bulletin of Science, Technology & Society*, 32 (5): 353–360. doi: 10.1177/0270467612469067.

Kronsell, A. (2012) *Gender, sex, and the postnational defense: militarism and peacekeeping.* Oxford studies in gender and international relations. New York: Oxford University Press.

Kunashakaran, S. (2016) Un(wo)manned aerial vehicles: an assessment of how unmanned aerial vehicles influence masculinity in the conflict arena. *Contemporary Security Policy*, 37 (1): 31–61. doi: 10.1080/13523260.2016.1154405.

Lambeth, Benjamin, S. (1997) The technology revolution in air warfare. *Survival*, 39 (1): 65–83.

Lee, J.-k. (2009) Surrogate military, subimperialism, and masculinity: South Korea in the Vietnam War, 1965–73. *Positions: East Asia Cultures Critique*, 17 (3): 655–682. doi: 10.1215/10679847-2009-019.

Liddell Hart, B.H. (1980) *The revolution in warfare.* Westport, CT: Greenwood Press.

Luttwak, E. (1995) Toward post heroic warfare. *Foreign Affairs*, 74 (3): 109.

MacKenzie, M.H. (2012) *Female soldiers in Sierra Leone: sex, security, and post-conflict development.* Gender and political violence series. New York: New York University Press.

Mahnken, T.G. (2010) *Technology and the American way of war since 1945*. New York; Chichester: Columbia University Press.

Manjikian, M. (2014) Becoming unmanned: the gendering of lethal autonomous warfare technology. *International Feminist Journal of Politics*, 16 (1): 48–65.

Marchand, M.H. (1998) Different communities/different realities/different encounters: a reply to J. Ann Tickner. *International Studies Quarterly*, 42 (1): 199–204. doi: 10.1111/0020-8833.00077.

Martin, S.B. (2002) The role of biological weapons in international politics: the real military revolution. *Journal of Strategic Studies*, 25 (1): 63–98. doi: 10.1080/714004040.

Marway, H. (2011) Scandalous subwomen and sublime superwomen: exploring portrayals of female suicide bombers' agency. *Journal of Global Ethics*, 7 (3): 221–240. doi: 10.1080/17449626.2011.635677.

Masters, C. (2005) Bodies of technology: cyborg soldiers and militarized masculinities. *International Feminist Journal of Politics*, 7 (1): 112–132. doi: 10.1080/1461674042000324718.

Masters, C. (2009) Femina sacra: the 'War on/of Terror', women and the feminine. *Security Dialogue*, 40 (1): 29–49. doi: 10.1177/0967010608100846.

Masters, C. (2010) Cyborg soldiers and militarised masculinities. *Eurozine*, 20 May. Available at: www.eurozine.com/articles/2010-05-20-masters-en.html (accessed: 3 July 2016).

Matthew, S. (2015) I'll BEE back: fleets of Terminator-style drones could have artificial brains based on honeybees. *Daily Mail*, 25 April. Available at: www.dailymail.co.uk/news/article-3055152/I-ll-BEE-Fleets-Terminator-style-drones-artificial-brains-based-honeybees.html (accessed: 3 July 2016).

Mayer, J. (2009) The Predator war: what are the risks of the CIA's covert drone program? *The New Yorker*, 26 October. Available at: www.newyorker.com/magazine/2009/10/26/the-predator-war (accessed: 14 November 2012).

McCarry, M. (2007) Masculinity studies and male violence: critique or collusion? *Women's Studies International Forum*, 30 (5): 404–415. doi: 10.1016/j.wsif.2007.07.006.

McKinley, R.A., McIntire, L.K. and Funke, M.A. (2011) Operator selection for unmanned aerial systems: comparing video game players and pilots. *Aviation, Space, and Environmental Medicine*, 82 (6): 635–642. doi: 10.3357/ASEM.2958.2011.

McMaster, H.R. (2008) Learning from contemporary conflicts to prepare for future war. *Orbis*, 52 (4): 564–584. doi: 10.1016/j.orbis.2008.07.011.

McNeill, W.H. (1993) *The pursuit of power: technology, armed force, and society since A.D. 1000*. 7. Nachdr. Chicago, IL: University of Chicago Press.

McSorley, K. (ed.) (2013) *War and the body: militarisation, practice and experience*. New York: Routledge.

Millar, Katharine M. and Tidy, Joanna (2017) Combat as a moving target: masculinities, the heroic soldier myth, and normative martial violence. *Critical Military Studies*, 3 (2): 142–160.

Mitchell, B. (1989) *Weak link: the feminization of the American military*. Washington, DC; New York: Regnery Gateway; Distributed to the trade by Kampmann.

Mohanty, C.T. (2003) *Feminism without borders: decolonizing theory, practicing solidarity*. Durham; London: Duke University Press.

Moon, K. (1999) South Korean movements against militarized sexual labor. *Asian Survey*, 39 (2): 310–327. doi: 10.2307/2645457.

Morgan, D. (1994) Theatre of war: combat, the military, and masculinities. In Brod, H. and Kaufman, M. (eds) *Theorizing masculinities*. London: Sage Publications. pp. 165–182.

NATO (2003) *Cyberwar-Netwar*. Advanced Research Workshop on Cyberwar-Netwar: Security in the Information Age. IOS Press.

Neuneck, G. (2008) The revolution in military affairs: its driving forces, elements, and complexity. *Complexity*, 14 (1): 50–61. doi: 10.1002/cplx.20236.

Newpower, A. (2006) *Iron men and tin fish: the race to build a better Torpedo during World War II*. Annapolis, MD: Naval Institute Press.

Oldenziel, R. and Zachmann, K. (eds) (2009) *Cold War kitchen: Americanization, technology, and European users*. Inside technology. Cambridge, MA: MIT Press.

Pagallo, U. (2011) Robots of just war: a legal perspective. *Philosophy & Technology*, 24 (3): 307–323. doi: 10.1007/s13347-011-0024-9.

Parker, G. (1996) *The military revolution: military innovation and the rise of the West, 1500–1800*. 2nd edn. Cambridge; New York: Cambridge University Press.

Peoples, C. (2010) *Critical security studies: an introduction*. Milton Park, Abingdon, Oxon; New York: Routledge.

Peterson, V.S. (1992) Transgressing boundaries: theories of knowledge, gender and international relations. *Millennium: Journal of International Studies*, 21 (2): 183–206.

Peterson, V.S. and Runyan, A.S. (2010) *Global gender issues in the new millennium*. Dilemmas in world politics. 3rd edn. Boulder, CO: Westview Press.

Pettman, J. (1996) *Worlding women a feminist international politics*. London; New York: Routledge. Available at: http://public.eblib.com/choice/publicfullrecord.aspx?p=254158 (accessed: 7 November 2014).

Pick, O., Kintner, W.R., and Scott, H.F. (1969) The nuclear revolution in Soviet military affairs. *International Affairs (Royal Institute of International Affairs 1944–)*, 45 (1): 151. doi: 10.2307/2612669.

Pin-Fat, Veronique and Stern, Maria (2005) The scripting of Private Jessica Lynch: biopolitics, gender, and the 'feminization' of the U.S. military. *Alternatives: Global, Local, Political*, 30 (1): 25–53.

Power, M. (2013) Confessions of a drone pilot. *GQ*, 23 October. Available at: www.gq.com/news-politics/big-issues/201311/drone-uav-pilot-assassination (accessed: 18 September 2014).

Puar, J.K. (2005) Queer times, queer assemblages. *Social Text*, 23 (3–4): 121–139.

Rajan, V.G.J. (2011) *Women suicide bombers: narratives of violence*. Critical terrorism studies. Milton Park, Abingdon, Oxon; New York: Routledge.

Ramazanoglu, C. (1992) What can you do with a man? *Women's Studies International Forum*, 15 (3): 339–350. doi: 10.1016/0277-5395(92)90002-D.

Rappert, B. and Moyes, R. (2009) The prohibition of cluster munitions. *The Nonproliferation Review*, 16 (2): 237–256. doi: 10.1080/10736700902969687.

Rayner, A. (1994) Cyborgs and replicants: on the boundaries. *Discourse: Studies in the Cultural Politics of Education*, 16 (3): 124–143.

Reichmann, F. (1945) The Pennsylvania rifle: a social interpretation of changing military techniques. *The Pennsylvania Magazine of History and Biography*, 69 (1): 3–14.

Richter-Montpetit, M. (2007) Empire, desire and violence: A queer transnational feminist reading of the prisoner 'abuse' in Abu Ghraib and the question of 'gender equality'. *International Feminist Journal of Politics*, 9 (1): 38–59. doi: 10.1080/14616740601066366.

Riza, M.S. (2013) *Killing without heart: limits on robotic warfare in an age of persistent conflict*. First edition. Washington, DC: Potomac Books.

Roff, H.M. (2014) The strategic robot problem: lethal autonomous weapons in war. *Journal of Military Ethics*, 13 (3): 211–227. doi: 10.1080/15027570.2014.975010.

Roff, H.M. (2016) Gendering a warbot: gender, sex and the implications for the future of war. *International Feminist Journal of Politics*, 1: 1–18. doi: 10.1080/14616742.2015.1094246.

Rogers, C.J. (ed.) (1995) *The military revolution debate: readings on the military transformation of early modern Europe*. History and warfare. Boulder, CO: Westview Press.

Rogers, C.J. (2011) The development of the longbow in late medieval England and 'technological determinism'. *Journal of Medieval History*, 37 (3): 321–341. doi: 10.1016/j.jmedhist.2011.06.002.

Royakkers, L. and van Est, R. (2010) The cubicle warrior: the marionette of digitalized warfare. *Ethics and Information Technology*, 12 (3): 289–296. doi: 10.1007/s10676-010-9240-8.

Sauer, F. and Schornig, N. (2012) Killer drones: the 'silver bullet' of democratic warfare? *Security Dialogue*, 43 (4): 363–380. doi: 10.1177/0967010612450207.

Scarry, E. (1987) *The body in pain: the making and unmaking of the world*. New York: Oxford University Press.

Schweitzer, Y. and Merkaz Yafeh le-meḥkarim asṭraṭegiyim (eds) (2006) *Female suicide bombers: dying for equality?* Memorandum no. 84. Tel Aviv, Israel: Jaffee Center for Strategic Studies, Tel Aviv University.

Sharkey, N. (2008) Cassandra or false prophet of doom: AI robots and war. *IEEE Intelligent Systems*, 23 (4): 14–17. doi: 10.1109/MIS.2008.60.

Sharkey, N. (2011) Automating warfare: lessons learned from the drones. *Journal of Law, Information & Science*, 21 (2): 140–154. doi: 10.5778/JLIS.2011.21.Sharkey.1.

Shepherd, L.J. (2006) Veiled references: constructions of gender in the Bush administration discourse on the attacks on Afghanistan post-9/11. *International Feminist Journal of Politics*, 8 (1): 19–41. doi: 10.1080/14616740500415425.

Shepherd, L.J. (2008a) *Gender, violence and security: discourse as practice*. London; New York: Zed Books; Distributed exclusively in the USA by Palgrave Macmillan.

Shepherd, L.J. (2008b) Visualising violence: legitimacy and authority in the 'War on Terror'. *Critical Studies on Terrorism*, 1 (2): 213–226. doi: 10.1080/17539150802184611.

Shepherd, L.J. (ed.) (2010) *Gender matters in global politics: a feminist introduction to international relations*. New York: Routledge.

Short (1999) The role of NGOs in the Ottawa process to ban landmines. *International Negotiation*, 4 (3): 483–502. doi: 10.1163/15718069920848589.

Simpson, T.W. (2011) Robots, trust and war. *Philosophy & Technology*, 24 (3): 325–337. doi: 10.1007/s13347-011-0030-y.

Singer, P.W. (2011) *Wired for war: the robotics revolution and conflict in the 21st century*. New York: Penguin.

Sjoberg, L. (2007) *Mothers, monsters, whores: women's violence in global politics*. New York: Zed Books.

Sjoberg, L. (2010a) Gendering the empire's soldiers: gender ideologies, the U.S. military, and the 'War on Terror'. In Sjoberg, L., Via, S., and Enloe, C. (eds) *Gender, war and militarism: feminist perspectives*. Santa Barbara, CA: Praeger. pp. 209–218.

Sjoberg, L. (2010b) Women fighters and the 'beautiful soul' narrative. *International Review of the Red Cross*, 92 (877): 53–68. doi: 10.1017/S181638311000010X.

Sjoberg, L. (2011) Looking forward, conceptualizing feminist security studies. *Politics & Gender*, 7 (04): 600–604. doi: 10.1017/S1743923X11000420.

Sjoberg, L. (2012) Gender, structure, and war: what Waltz couldn't see. *International Theory*, 4 (01): 1–38. doi: 10.1017/S175297191100025X.

Sloan, E.C. (2002) *The revolution in military affairs: implications for Canada and NATO*. Foreign policy, security and strategic studies. Montreal: McGill-Queen's University Press.

Smith, C.M. (2016) Gaze in the military: authorial agency and cinematic spectatorship in 'drone documentaries' from Iraq. *Continuum*, 30 (1): 89–99. doi: 10.1080/10304312. 2015.1117571.

Smith, J. (1993) *Misogynies: reflections on myths and malice.* New and rev. ed. London: Faber and Faber.

Sparrow, R. (2007) Killer robots. *Journal of Applied Philosophy*, 24 (1): 62–77. doi: 10.1111/j.1468-5930.2007.00346.x.

Sparrow, R. (2016) Robots and respect: assessing the case against autonomous weapon systems. *Ethics & International Affairs*, 30 (01): 93–116. doi: 10.1017/ S0892679415000647.

Stachowitsch, S. (2012) Military gender integration and foreign policy in the United States: a feminist international relations perspective. *Security Dialogue*, 43 (4): 305–321. doi: 10.1177/0967010612451482.

Steans, J. (2003) Engaging from the margins: feminist encounters with the 'mainstream' of international relations. *The British Journal of Politics and International Relations*, 5 (3): 428–454. doi: 10.1111/1467-856X.00114.

Steans, J. (2013) *Gender and international relations.* Cambridge, UK : Malden, MA: Polity Press.

Stone, J. (2004) Politics, technology and the revolution in military affairs. *Journal of Strategic Studies*, 27 (3): 408–427. doi: 10.1080/1362369042000282967.

Svechin, A. and Lee, K.D. (2004) *Strategy.* Minneapolis, MN: East View Publications.

Sylvester, C. (1994) *Feminist theory and international relations in a postmodern era.* Cambridge studies in international relations 32. Cambridge, UK; New York: Cambridge University Press.

Sylvester, C. (2013) *War as experience: contributions from international relations and feminist analysis.* War, politics and experience. Milton Park, Abingdon, Oxon; New York: Routledge.

Taylor, Diana (1993) Spectacular bodies: gender, terror, and Argentina's 'Dirty War'. In Cooke, Miriam and Woollacott, Angela (eds) *Gendering war talk.* Princeton, NJ: Princeton University Press. pp. 20–42.

The Economist (2012) March of the robots. *The Economist*, 403 (8787).

Tickner, J.A. (1992) *Gender in international relations: feminist perspectives on achieving global security.* New York: Columbia University Press.

Tickner, J.A. (1997) You just don't understand: troubled engagements between feminists and IR theorists. *International Studies Quarterly*, 41 (4): 611–632.

Travers, T. (1992) *How the war was won: command and technology in the British Army on the western front, 1917–1918.* London; New York: Routledge.

United Nations Development Programme (ed.) (1994) *Human development report 1994.* New York: Oxford University Press.

Vallor, S. (2013) *The future of military virtue: autonomous systems and the moral deskilling of the military.* Published in: 2013 5th International Conference on Cyber Conflict (CYCON 2013). Date of conference: 4–7 June 2013, Tallinn, Estonia.

Van Creveld, M. (1991) *Technology and war: from 2000 BC to the present.* A rev. and expanded ed., 1st Free Press ed.; 1st Free Press paperback edn. New York; Toronto: Free Press; Maxwell Macmillan Canada; Maxwell Macmillan International.

Van Creveld, M. (2001) *Men, women, and war.* London: Cassell & Co.

Van Creveld, M. (2013) To wreck a military. *Small Wars Journal.* Available at: http:// smallwarsjournal.com/jrnl/art/to-wreck-a-military (accessed: 13 October 2015).

Wall, T. and Monahan, T. (2011) Surveillance and violence from afar: the politics of drones and liminal security-scapes. *Theoretical Criminology*, 15 (3): 239–254. doi: 10.1177/1362480610396650.

Weber, C. (1994) Good girls, little girls, and bad girls: male paranoia in Robert Keohane's critique of feminist international relations. *Millennium – Journal of International Studies*, 23 (2): 337–349. doi: 10.1177/03058298940230021401.

Whitworth, S. (2005) Militarized masculinities and the politics of peacekeeping: the Canadian case. In Booth, K. (ed.) *Critical security studies in world politics*. Boulder, CO: Lynne Rienner Publishers. pp. 89–106.

Wilcox, L.B. (2015) *Bodies of violence: theorizing embodied subjects in international relations*. Oxford Studies in gender and international relations. Oxford; New York: Oxford University Press.

Wilcox, Lauren (2017) Embodying algorithmic war: gender, race, and the posthuman in drone warfare. *Security Dialogue*, 48 (1): 11–28.

Williams, P. (ed.) (2008) *Security studies: an introduction*. London; New York: Routledge.

van Woudenberg, N. (2008) The long and winding road towards an instrument on cluster munitions. *Journal of Conflict and Security Law*, 12 (3): 447–483. doi: 10.1093/jcsl/krn007.

Zalewski, M. (1993) Feminist standpoint theory meets international relations theory: a feminist version of David and Goliath. *Fletcher Forum of World Affairs*, 17 (2): 13–32.

Zalewski, M. (1995) Well, what is the feminist perspective on Bosnia? *International Affairs*, 71 (2): 339–356.

Zalewski, M. (2000) *Feminism after postmodernism: theorising through practice*. London; New York: Routledge.

Zalewski, M. and Parpart, J.L. (eds) (1998) *The 'man question' in international relations*. Boulder, CO: Westview Press.

2   Haunting

Introduction: 'life is complicated'[1]

Feminist security scholars have explored the interaction between gender and military technologies, but to date their theorising has tended to view these technologies as *either* masculinising *or* feminising. As noted in the previous chapter, these narratives are not able to contain or explain the contradictions that are woven through the data on military technologies, and in particular, the debates surrounding the use of drones. What I want to add to the feminist scholarship on drones, and military technologies more generally, is a framework that provides a means through which to better capture the nuance, detail and complexity of the phenomena under analysis. To do this I build on the work of Avery Gordon (2008), Jacques Derrida (2006) and Jessica Auchter (2014) and their frameworks of 'Haunting'.[2] And as such, will, in this chapter explore how

> Haunting is an encounter in which you touch the ghost or the ghostly matter of things: the ambiguities, the complexities of power and personhood, the violence and the hope, the looming and receding actualities, the shadows of ourselves and our society.
>
> (Gordon, 2008, p. 134)

This requires taking a particular stance on what it means to speak about ghosts, about traces, about the way that gender can be conceived of as a spectral structure of power. In so doing, I argue that using Haunting as a framework enables me to situate the narratives of gender and military technologies within the complex (cob)webs of power in which they function.[3]

In the first part of this chapter I contextualise Haunting within the 'spectral (re)turn' exploring the different ways in which Haunting and its component of the ghost have been deployed. I argue that complexity and nuance are foregrounded by Haunting through four core themes: complex personhood, in/(hyper)visibility, disturbed temporality, and power, which I sketch out in detail.

The second part of this chapter outlines the ghost hunt, the methodology of Haunting. The ghost hunt provides a means of exploring how to take seriously the data which does not 'fit' (e.g. neatly into the category of *either* masculine *or* feminine) and how to engage with 'sensuous' and 'deviant' knowledges (concepts I return to) (Gordon, 2008, p. 205; Ahmed, 2006). As Holloway and Kneale note,

> the power of ghosts [is the] contingent potentiality of haunted spaces to enchant, to make us wonder and in so doing usher further new interpretative frameworks, to open up our epistemologies or even our ontologies in fantastic, strange and sometimes baffling ways....
>
> (2008, p. 308)

This chapter will close by detailing the way in which I augment the 'ghost hunt' with the language of Cynthia Weber's (2014, 2016) 'queer logic(s)'. I do this because, whilst Haunting is already concerned with the 'things that don't fit' and with the way that ghosts *are the same time* present and absent, I argue that the addition of queer logic(s) provides the necessary vocabulary to render this binary destabilisation explicit. In keeping with the concerns of the poststructural feminist underpinnings of the book, I utilise queer logic(s) to express the ways in which military technologies *simultaneously* destabilise *and* (re)inscribe masculinity and femininity. To date, as my outline of the 'spectral (re)turn' will indicate, uses of Haunting and similar work has tended to be focused around literary criticism, sociology, and history. And whilst I am not the first to implement the framework in relation to the concerns of International Relations, I hope that the addition of queer logic(s) to the language of the ghost hunt will make clearer the usefulness of this perspective to International Relations concerned with the (cob)webs of power woven out of intersectional identifiers like gender.

The spectral (re)turn[4]

Haunting is part of the wider 'spectral (re)turn' in social science and has been applied to a wide range of subject areas including sociology (Gordon, 2008, 1990; Roseneil, 2009), International Relations (Auchter, 2014; Welland, 2013), videography (Garrett, 2011; Simpson, 2011), art (Hawkins, 2010; Tolia-Kelly, 2004), cartography (Cosgrove, 2008; Jacob and Dahl, 2006), literature (Luckhurst, 2002), history (Kleinberg, 2017), city planning (Comaroff, 2007; Edensor, 2005) and geography (Mayerfield Bell, 1997; Roberts, 2013). What these various perspectives share is an interest is the 'stuff' that is *more than* the visible, the concrete, the objective.[5]

In a move attributed by many to Jacques Derrida (2006), Nicholas Abraham, and Maria Torok (1999; 2008), the 1990s saw the rehabilitation of 'ghosts' from 'possible actual entities, plot devices and clichés of common parlance' (and as such, inappropriate subjects for serious academic investigation); to 'conceptual metaphors' (Blanco and Peeren, 2013, p. 1). As a 'conceptual metaphor' the ghost acts as a 'discourse, a system of producing knowledge' providing insight into areas previously passed over or understudied (2013, p. 1). Despite an interest in 'ghosts' (figures I will return to shortly), Haunting is not concerned with the occult or with parapsychology, but rather requires paying attention to the unsaid, the absent, and silent just as much as the spoken, the present, and the evident. Gordon's version of Haunting,[6] which inspires the framework used here, pays attention to these neglected forms of knowledge as a way of engaging with the claim that 'life is complicated' on both a practical and theoretical basis and I use this statement as the start point for my theorising of the way that drones may be viewed as simultaneously destabilising and (re)inscribing military masculinity.[7]

Avery Gordon's (1990, 2008) use of Haunting drew me to consider the ghostly, and folds together two of her primary concerns: first, she is interested in how structures of power such as what we call 'racism' and 'sexism' are experienced, how these terms imply something concrete that is also nebulous, something personal which is at the same time impersonal. Second, in her investigation of the first, she develops Haunting as an attempt to go beyond what she views as the methodological failings of sociology as a social science. In order to investigate these two issues, Gordon's *Ghostly matters* provides a meditation on how 'being haunted' has been and might be understood beginning with a discussion of Freud and the development of psychoanalysis as a field. Therefore, what follows is a brief outline of its relevance to the ghostly and my own framework.

Freud was 'unsure' about ghosts. His work (1997) indicates that he viewed himself as an educated man who had moved beyond the pre-modern era of superstition, and yet his theorising returns again and again to the ghostly. Gordon (2008) is attracted to Freud and psychoanalysis for three primary reasons. First, because the ghost who initially piques her interest is associated with the field. (The ghostly woman, Sabina Spielrein, is a key figure who we will return to.) Second, whilst Gordon is interested in Freud's conceptualisation of the unconscious, she is critical of his determination to ignore the social and his focus on 'enclosing it within the subject' (2008, p. 48). Third, Freud's 'uncanny' provides the vehicle through which she draws attention to 'qualities of feeling' and her idea of sensuous knowledge(s) (2008, p. 50). We will go through a more detailed exploration of Gordon's analysis in a moment, but the end point, she notes, is that 'Freud will disappoint', marking the point from which she departs from psychoanalysis as a methodological tool (2008, p. 42).

First then, I want to introduce you to the character of Sabina Spielrein. Actually, the term 'character' is a misnomer, Spielrein was a 'real' historical figure, a person, but the term character is something that I want to go on to unpick.[8] Briefly then, Spielrein is a young Jewish girl who, developing schizophrenia, is taken to the psychiatric clinic run by Carl Jung. Here she is 'successfully' treated and becomes a student of psychoanalysis. However, Spielrein also embarks on a love affair with Jung, which he breaks off when he fears a scandal. Both Spielrein and Jung confide in Freud, and their experiences inform the development of various theories that later form part of the canon of psychoanalysis. Gordon is drawn to Spielrein's story by a photograph of the participants of the Third Psychoanalytic Congress held in Weimar in 1911. It is this photo in which Gordon notices Spielrein's absence. Spielrein's absence from this photograph leads Gordon to trace Spielrein's absence from the theories which she helped develop, most noticeably regarding the 'Death Drive'.[9] As such, Gordon draws attention to the (cob)webs of power which might be called 'sexism' and the way that they are woven not just with historical fact, but with hopes, desires, and discipline; with paranoia and pain. This chapter will revisit the experiences of Spielrein and Gordon's effort to engage with her (ghost), but the lack of Spielrein, her absence (in a photo initially) in the body of psychoanalysis literature is what draws Gordon into engaging with Freud and his complex relationship with the ghostly.

Gordon's second and third point of interest in psychoanalysis results from Freud's apparent desire to avoid dealing with the ghostly, whilst continually returning to it. To explore this in more detail it is necessary to briefly introduce the concept of 'the uncanny' (Freud, 1997). To experience the uncanny is to experience the recognisable and well-known rendered strange: 'a familiar room become alien' (Gordon, 2008, p. 50). This horror of the blending of the line of distinction between familiar and strange is something that deeply interests Gordon, and that she claims is 'not simply ... cognitive doubt, or ... the unknown, but something else' (2008, p. 31). She is, therefore, frustrated by Freud's refusal to acknowledge or engage with the ghostly substance of this 'something else' when he stipulates that the uncanny is nothing more than 'repressed infantile complexes' or 'primitive beliefs' (Freud cited in Gordon, 2008, p. 52). Indeed, Freud goes on to collapse the two categories into one another so that the 'primitive beliefs' that he points to amount to nothing more interesting or social than a manifestation of those 'repressed infantile complexes'. Therefore, the ghostly, the uncanny, as Freud appears to view it is little more than a symptom of underlying psychosis, or part of the unconscious of the individual. What Gordon makes of this 'something else' I introduce below, but in relation to her thinking on Freud, she notes disappointedly that 'Freud gets so close to dealing with the social reality of haunting only to give up the ghost and everything social that comes in its wake' (2008, p. 57).

Freud's work and the wider canon of psychoanalysis then informs the position from which Gordon, and I, begin to think about Haunting. Gordon's analysis of Freud points to several areas in which his work fails to account for the complexity of Haunting, most notably in his refusal to engage with the social workings of the ghost. Through his (and Jung's) failure to acknowledge the scholarship of their former patient Spielrein, Freud inadvertently draws attention to his failure to adequately engage with spectral structures of power (such as sexism) which are implicated in the experience of being haunted.

Outside of psychoanalysis, but informed by the topic, ghosts and the ghostly are explored in order to investigate trauma and memory. Jessica Auchter's (2014) *The Politics of Haunting and Memory in International Relations*, for example, traces the ways in which certain spaces are haunted by the deaths and/or memorialisation that occurred there. Auchter's work approaches Haunting from a slightly different perspective from Gordon, developing a 'hauntology' as an alternative to ontology.[10] She investigates how statecraft relies on the construction of the distinction between life and death in order to better understand the process of memorialisation. Like Gordon and myself, Auchter is troubled by the lines drawn between categories, specifically the categories of life and death and presence and absence, and seeks to use Haunting, and the ghosts (who are neither alive nor dead) to 'trouble the very existence of that dichotomy' (2014, p. 5). However, Auchter does not begin her thinking about Haunting from psychoanalysis but rather begins by engaging with Derrida's (2006) *Specters of Marx*. Taking Derrida's primary concern to be the illumination of 'the politics behind the construction of dichotomies at an ontological level' Auchter focuses on how this plays out in places/spaces of death and memorialisation (2014, p. 4). These spaces haunt her and, like Gordon, she points to the affective nature of reckoning with, speaking with, ghosts.

Derrida's (2006) *Specters of Marx* is the primary text outside of Freud's works to which Haunting is attributed. The book builds on Derrida's remarks at a conference about the potential future for Marxism after the fall of the Berlin Wall, noting that there were many 'specters' which needed to be taken seriously by scholars at that time. Important for our discussion here is Derrida's interest in the way that 'time is out of joint' (drawn from Shakespeare's *Hamlet*) and that there are plural spectres of Marx: 'Why this plural? Would there be more than one of them? Plus d' un [More than one/No more one]' (2006, pp. 1–2). The first of these feeds into what I term the thematic concern of disturbed temporality and which I outline in detail in a later section. The latter, gestures (again) to the recurring theme of binary destabilisation and the way in which the ghostly invites a questioning of the boundary line between dyads such as presence/absence, life/death, masculine/feminine. In order to question this boundary both Gordon and Auchter draw from Derrida's writing a particular way of dealing with ghosts ('There is then some spirit. Spirits. And one must reckon with them.'), specifically that one must be 'hospitable' to them (Derrida, 2006, pp. xx, 135).

Both Auchter and Gordon take seriously Derrida's injunction to be hospitable to ghosts, acknowledging an ethical commitment that comes from engaging with the spectral: 'it is in the name of justice' (Derrida, 2006, p. xviii). For Auchter, being hospitable to ghosts means exploring how the process of Othering, of other-ness is implicated in the way that the apparatus of the state (broadly conceived) constructs the line between life and death: who gets to live, who to die, whose death is marked and whose is 'ungrievable' (after Butler, 2004a). For Gordon, hospitality means following the traces of what is labelled, for example, racism or sexism and conveying the delicate but robust ways in which the lived reality of those labels function. Similarly, it is this ethical commitment that I engage with through my own framework of Haunting. But perhaps before discussing in any more detail what it means to *engage* with ghosts it might be useful to explore what precisely I mean when I speak or write about ghosts.

Gespensts, revenants and spuk (ghosts)

Ghosts have had a varied past,

> appearing as anything from figments of the imagination, divine messengers, benign or exacting ancestors, and pesky otherworldly creatures populating

particular loci to disturbing figures returned from the dead bent on exacting revenge, revealing hidden crimes, continuing a love affair or simply searching for a way to pass on.

(Blanco and Peeren, 2013, p. 1)

Ghosts are unsurprisingly a core component of every Haunting framework, but how these entities are understood and utilised varies widely across the literature. The main distinctions focus on questions of what they are, whether they should be understood literally or metaphorically, what work they do (if any), and how we can/should/ought to respond to them.

Psychologists Abraham and Torok (1999; 2008) sought ghosts who represented undisclosed transgenerational secrets, 'a dead ancestor in the living Ego' (Davis, 2005, p. 374). Therefore, they viewed the ghostly as difficult to access but once reaching the 'curative moment', as something that could be resolved. Despite writing the foreword for Abraham and Torok's texts, Derrida's understanding of the ghost, which informs my own, is very different. Derrida's specters are not frightening psychological disturbances, but rather represent that which stands 'in defiance of binary oppositions' (Buse and Stott, 1999, p. 10). Being neither dead *nor* alive, being simultaneously present *and* absent, the ghostly points to the inherent instability of these dyads in ways that are particularly useful to thinking about masculinity and femininity.

Therefore, where Freud (1997) and Adorno (1974) fought against the conceptual figure of the ghost as a crass reminder of pre-modern superstitions and a lack of scientific knowledge, Derrida (2006) sought instead to rehabilitate the ghost through a deconstructive re-reading of Marx. As a result, Derrida's ghosts resist resolution. From his perspective, rather than representing either primitive/pre-enlightenment thinking or something frightening, the ghost is a figure that needs to be heard instead of being exorcised. The things that the ghost has to say are important because, according to Derrida, they indicate 'the structure of every hegemony' and the conversations with ghosts are 'a matter of neutralizing a hegemony or overturning some power' (Derrida, 2006, pp. 46, 58).

To think about how to engage with ghosts as a means of 'overturning some power' it is necessary to situate Haunting within the wider body of social science scholarship. One of the things that unites many, if not all, of the uses of Haunting, is a concern that positivist forms of research are missing important things. This perspective, in aiming for an elusive (impossible?) 'objective' stance, results in a lack of attention being paid to the political implications of epistemological and ontological choices: 'blinding us to the ways in which those things are expressly produced and fundamentally enabled by a history of loss and repression' (Janice Radway's preface to Gordon, 2008, p. ix). The immense canon of *post*modern scholarship that developed in response to this concern is one Gordon engages with noting that 'How I came to write a book about ghostly matters is a long story and some of that story has to do with postmodernism' (2008, p. 8). Whilst critical of 'social science's empiricist grounds for knowing', Gordon is also uncomfortable with the resulting response, as she see it, from postmodern critiques (2008, p. 9). In relation to postmodernism, she is frustrated by the continuing salience of binaries. She is frustrated by the apparent inability of postmodernism to handle the distinction between 'reality and its modes of production' and what she views as the failure of postmodernism to *address* the hierarchical structures which the perspective necessarily draws attention to (Gordon, 2008, p. 11).

Gordon's frustrations are of particular interest to a feminist poststructuralist. One of the core concerns of feminist scholarship during the development of post-modern and post-structural thinking was whether or not these perspectives could be compatible with an emancipatory feminist agenda. Can the political aims of feminism ever sit comfortably alongside an ontological and epistemological stance that questioned the existence of a person called 'woman'? How can fem-inists work to 'overturn some power', challenge the structures of patriarchy, if all that was left of 'reality' was representation? As I outlined in the previous chapter, it is entirely possible for postmodern/post-structural feminism to exist, but Gordon's discomfort with 'the strange new landscape' of postmodernity is instructive for thinking about how she uses Haunting as an alternative, and, in particular, how it relates to what and how we (as social scientists) understand or construct 'truth' (Jameson cited in Gordon, 2008, p. 12).

I am anxious that the 'truth' that I posit in this book be understood as one of other possible 'truths', whilst also attempting to avoid lapsing into relativism and fiction. As Kleinberg (2017) notes, there is a problem with referring to the past as a fixed entity because the present is always becoming the past, so there is an ever increasing, ever changing nature to that which we label 'past'. Therefore, even though my project deals with contemporary issues, these too are already trickling into history. The unfixity of time is in direct contrast, Kleinberg argues, with a positivist Aristotelian conceptualisation of the truth that rests in 'the realm of the eternally same' (2017, p. 6). Therefore, seeking to rest a conceptualisation of the truth on an account that claims to be 'what really happened' is a danger-ous venture. What I take from this, and weave through my understanding of Gordon's framework is that there is a tension to be negotiated between an inse-cure and unsettled claim to 'truth', the reflection of 'truth' in reality (however understood), and in refusing the teleological endpoint implied by arguments about progress.[11] Therefore, to avoid the twin pitfalls of relying on metanarra-tives and the prescriptions of Modernist thinking it is necessary to pay attention to the particular, the detail, the abstruse, and the intricate. By focusing on these we can better understand what is signified when we use terms like 'sexism' and 'racism', and we can better understand how the spectral structures of power in those terms have lived and disciplinary consequences.

The ghostly, then, in its capacity to stand 'in defiance of binary oppositions' not only illuminates the spectral structures of power behind terms like racism and sexism but *also* points to their constructed nature, forces a questioning of those inequalities often defended under the guise of being 'natural' (Buse and Stott, 1999, p. 10). Haunting brings together the personal and subjective with discourses that dis/empower (gender, sex, race, class, sexuality as well as culture, history and society), braiding together the threads which reflect the construction of social life. Therefore, I argue that Haunting is profoundly important for feminist theorising, providing a means of illuminating the cobwebbed structure of patriarchy and, perhaps, pointing to some ways of 'overturning some power' in the pursuit of a more equal society (Derrida, 2006, p. 58).

Inspired by the different possible readings, I understand 'ghosts' as something to be engaged with rather than eradicated or resolved, indeed I understand ghosts as inherently irresolvable, not as a puzzle or a problem. Similarly, I understand the ghost as a conceptual metaphor rather than as a purely linguistic device or representative of the occult/paranormal. This means that to engage with ghosts I am required, to paraphrase Donna Haraway (2017), to sit with the trouble. Sitting with the trouble in the context of gender and military technologies means starting from a point of discomfort (Why does this debate about drones not make sense? Why do the existing feminist engagements feel like they do not 'fit'? How can technology *simultaneously* masculinise *and* feminise?); and proceeding from those points of discomfort to ask questions and look for ghostly traces of what is not there. It means asking 'How and why does the myth of the duel continue to effect military subjectivity?' 'What does it mean to speak about being a "warrior" in the contemporary context and why is this contested?' 'Why does this contestation matter?' Using the thematic strands outlined in this chapter I point to specific ghosts, that is specific conceptual metaphors, which help me not to stabilise or reveal/resolve the issues but draw attention to the complexity and interweaving of social/political and cultural nuances at the intersection of gender and drones.

The understanding of ghosts that I deploy in this book does not just look at the ghostly as either the indicators of the distinction between life and death (although this is important), nor does it focus only on trauma and injustice. Rather, I explore the ghostly to draw attention to places and spaces of instability, contestation and discomfort. Drawing on Giorgio Agamben (1998) and Judith Butler (2004a), Auchter (2014) argues that Haunting is essential to understanding the way in which identity is politicised, the way some lives are grievable and others are not. In this book I go a step further, arguing that Haunting allows us to investigate the way in which subjectivity is not just created/destroyed in death but also as lives are lived. Esther Peeren (2014) has done some important work in this vein, illuminating how undocumented migrants live lives that are ghostly – not just in how they are constructed in discourses but by existing, she argues, in the shadows of what constitutes 'life'. Navigating a thin line between illuminating the uncertainties that make the lives of her subjects ('undocumented migrants, servants or domestic workers, mediums and missing persons') ghostly and denying their agency, Peeren asks how associations with the ghostly impact on their lives (2014, p. 5). Ultimately, she notes, 'In the cases discussed here, the ghost is a metaphor certain people (are made to) live as', which is not the same as arguing that this is who they *are* (if such a self can be said to pre-figure the social context in which the subject functions) (2014, p. 5). What Peeren illustrates is that the ghostly is not *always* about the already dead nor about living but rather the distinction between life and death and instability between those two possibilities.

One of the ways in which my work departs from Peeren (and Gordon and Auchter) is in who figures as its primary subjects. By choosing to concentrate on the lives of British Reaper crews I focus on the haunting that occurs in the lives of those individuals who exist, theoretically at least, in a position of strength. The Reaper crews are part of a powerful military, they are part of a long and rich social, cultural fabric, possessed of the power to legitimately wield violence – to, apparently, decide who lives and who dies. Therefore, where Haunting has been usefully applied to draw attention to those individuals who have been neglected by traditional security studies research, this book, in part, investigates the utility of Haunting as a methodological framework for thinking about those individuals who *appear* to be situated in a position of strength – at least partly because it helps to reveal the instability of that position of power and its relationship with/co-constitution through gender.

The figure of the ghost has a rich history (and future!) that reflects the various ways in which Haunting has been used in the social sciences. Moving beyond an investigation of the paranormal the term 'ghost' is used to refer to the figures who indicate the things that other methodological frameworks have missed. Having outlined where my understanding of the ghost sits within the spectral (re)turn the following section goes on to explain how the ghostly can form a framework. I do so by outlining four key thematic areas to my framework which I have labelled: complex personhood, in/(hyper)visibility, disturbed temporality, and power.

Complex personhood

Gordon's rationale for Haunting is Patricia Williams' statement that 'Life is complicated', and from this she argues that 'Complex personhood is the second dimension' of that theoretical statement (2008, pp. 4–5).[12] Taking personhood seriously is important because 'all people (albeit in specific forms whose specificity is sometimes everything) remember and forget, are beset by contradiction, and recognize and misrecognize themselves and others' (Gordon, 2008, p. 4). Williams and Gordon argue that by neglecting complex personhood, social science reifies categories that cannot hope to do justice to the lived reality of the lives of those individuals who make up the body (and bodies) of security studies. To do justice to the individuals whose lives are the subject of this book means understanding those individuals as complex social, psychological, cultural, and historical beings tangled in multiple (cob)webs of power that dis/empower in different ways that affect agency. As noted in the introductory chapter to the book, whilst the shared experience of piloting armed drones is what unites all of the individuals I interviewed, this is merely one thread of each of their lives – one line on the CV, one part of who they see themselves as. Recognising this connection and divergence is an important part of utilising Haunting.

Perhaps the most useful way of understanding how complex personhood can be conceptually useful is through the idea of 'becoming'.[13] A core contention of poststructural feminists, like Judith Butler (1999), is that sex (just as much as gender) is not something we are born with, but rather something which we must constantly (re)produce. As illustrated in the previous chapter, this perspective, whilst it unsettles the concept of a singular 'woman', helps us to articulate the way in which 'how women are represented (constructed) in language is a seriously political act' (Zalewski, 2000, p. 70). As such, if subjecthood, personhood, is always in the process of being constructed, if there is 'neither origin nor end' to what subjectivity can and should be, then there is space for change, to be different, to be otherwise (Salih, 2002, p. 3; see also Butler, 1999, 2004b; and Braidotti, 2002). This is not to imply that subjectivity is simply a matter of shrugging on a different 'self' each day, but rather that even those moments where what it means to be a specific individual appears to have 'congealed' the '"congealing" is itself an insistent and insidious practice, sustained and regulated by various social means' (Butler, 1999, p. 33). Therefore, whilst Butler was primarily concerned about the way in which sex and gender 'congealed', and I am too, I am *also* interested in the way that all the *other* pieces, threads, and sediments of what go into making a person congeal.

Gordon points to the multifaceted way in which what we call 'racism' is experienced by the characters in Toni Morrison's *Beloved*: through '*the nature of white man's work*, and the dialectics of violence and hatred ... [through] *disappointment ... the weather ... deformed feet and lost teeth ... [and] furniture without memories ...*' (2008, p. 4 (emphasis in the original)). All of these things, that make up the racism the characters experienced, are woven from the threads of what congealed to make *a white man* or *a black woman*. In addition to being what the characters experienced, these things are the means through which the characters' personhood, as complex as it is, 'becomes'. Earlier I introduced the 'character' of Sabina Spielrein. This real-life woman, who is no longer alive 'becomes' a 'character' through Gordon's reading of her. Gordon reads her biography and excerpts from her diary, notices her absence in a photograph, and write letters to Spielrein that this long dead woman cannot possibly read.

> Dear Sabina, I am uneasy about using your story, or the story of the places you were between, as a pretext for speaking about methodology and other matters, about needing or seeming to need a dead woman to enliven matters.... Is this why you have come back to haunt me, because the rumours of your recovery have reached you?
>
> (Gordon, 2008, p. 59)

When I first wrote about this I argued that the identity of Spielrein was fabricated/made real through Gordon's process, that the ghost was rendered visible through the layers and layers of semi-translucent 'facts' drawn together as a kind of collage. A colleague pointed out to me that through this understanding the 'the identity of Spielrein is "made real" in accordance with Gordon's imagination ... constructing in her own mind's eye a woman she never knew for her [Gordon's] own purposes ...', a proposal that made my colleague understandably uncomfortable.[14]

The idea of 'becoming' can, then usefully be employed to trace Spielrein's ghost in a subtly different way to that proposed by Butler, Braidotti, and the theorists of the previous section. A person's subjectivity may 'become' through the 'doing' of certain things associated with particular intersectional markers: 'doing' womanhood, 'doing' whiteness, etc. By exploring this through Haunting and the notion of complex personhood it is possible to see the way subjectivity is *also* made through the 'doing' of others. Becoming, in part, through the 'doing of others' means taking seriously the cultural commentary in which an identity is situated, means tracing the shapes of the lived experiences of what is called 'sexism' or 'racism', means looking at why the figure of the warrior raises its spectral head in accounts of modern technologically advanced warfare, and so on. This is not to deny individual agency nor voice, but rather to acknowledge the social situatedness in which every individual functions. Personhood is never fixed, even after death, but 'becomes' through the process of remembering and forgetting (by others), the things we see, the things we don't, the things that we notice are absent.

I attempt to address complex personhood in my thinking about the intersection between gender and drones by collating information from a range of different sources. Whilst I begin from interview data from the Reaper crews themselves, I have augmented this with cartoons drawn of drone pilots, with parliamentary debates, with the history/myth of the fighter pilot and other bits and pieces. Through these various components I draw attention to the way that the context in which the crews function is riven through with (gendered) inconsistencies, the way the identity of 'Reaper pilot/sensor' 'becomes' through their accounts and those of others, the way in which the lives inform the debate, the debate informs the lives, the way that pop cultural references inform, enrage, and aggravate.

Following the tenants of feminist methodological theorising that are espoused by Haunting, I have tried to remain reflexive throughout the drafting of this book. I have tried to be critical about the way that I am using the information out of which I create my haunted collage, be attentive to the ghostly traces that I am choosing to highlight (and, importantly, in thinking about the ghosts that I have *not* engaged with). My authorship and my capacity to comment on the lives of the individuals who form my case study is necessarily (also) always becoming. Whilst I have endeavoured to include as much information as possible, and to speak to as many individuals as I could, the data I have collected is, of course, incomplete. I am conscious that in writing what I have written I have created an account that others might use to say, 'this is so'. I therefore invite the reader to engage with this account critically and reflect as Kleinberg does on archival research when he notes 'Every archive is necessarily incomplete, and any addition to it in the future changes the past' (2017, p. 9). My research is not archival, but it does dip into the past, does dip into the future, and swirls this all together with the present in an attempt to draw attention to the importance of paying attention to the complexity of personhood and the way that this can help us to better understand the intersection between gender and military technology.

In/(hyper)visibility

Given the nature of the ghostly as something indistinct it is perhaps unsurprising that what I have termed, 'in/(hyper)visibility' is the second element of my framework of Haunting. This neologism of sorts captures the breadth and inflection of the myriad different ways in which it is possible to be (some kind of) invisible, and speaks to the ways in which what might be visible might not be immediately understandable, might be illegible. This section is concerned with the visible, the invisible, the hypervisible, the things that are somewhere between these categories and the ways even the unseen might be made known.[15] I have perhaps been a little linguistically promiscuous in my use of the term '*in*visible'. Strictly speaking, this means something of which there is *no* visible hint, however, throughout this book I will use the term invisible to *also* refer to those things of which there is a hint, a trace. I do so because the ghostly destabilises the boundary between the visible and the invisible – we *may* see something of the ghost or we may become aware of, come to 'know', its presence without visual cues. There are two ways that I use Haunting to approach in/(hyper)visibility, first through a consideration of what *is* (barely) visible but which is ignored; and second, through a consideration of what is felt/known but not seen (even as a trace). Core questions used in many academic engagements with Haunting are, 'What is it that we are *not* seeing?' 'Who is it that is *not* there?' And from these questions follows the methodological concern: 'How can we see what is not there?' As such, what the ghostly does is to 'muddle' with our ideas of what visibility is. I use this term after Haraway who explains that 'muddle' comes from the 'Old Dutch for muddying the waters' and which she uses 'to trouble the trope of visual clarity as the only sense of affect for mortal thinking' (2017, p. 174). In so troubling that trope, Haunting and the ghostly invite an exploration of the 'other-than' visual, the affective, what Gordon refers to as 'sensuous' knowledge(s) (2008, p. 205).

The analytical importance of engaging with in/(hyper)visibility is particularly evident in a consideration of its political implications. It matters what we 'see' because we have been taught that what we see is what *matters*, that the visible is what we should concern ourselves with. Feminist and queer scholars have explored this in relation to women and queer individuals, arguing that through rendering these individuals invisible (either by relegating them to the 'private' sphere or by refusing to speak/write their experiences), it has been possible to dismiss them as *not* mattering.[16] Therefore, by asking what we are *not* seeing, what is not visible, Haunting enables us to draw attention to the ways in which certain populations, figures or subjectivities have been marginalised or rendered unintelligible. In addition, feminist and queer theorists have shown how invisible subjects can simultaneously be rendered *hyper*visible. In Sarah Lamble's (2009) assessment of the *in*visibility of queer/gay/lesbian/trans- bodies in legal proceedings, she argues that queer/trans- individuals are rendered hypervisible by the absence of embodied and experiential references (as such, 'entrenched legal rationalities … [render] … queer bodies and sexualities unthinkable and unknowable') whilst detailed descriptions of what the individuals in questions were wearing, what they looked like, what activities they were engaged in render their 'Otherness' hypervisible (Lamble, 2009, p. 117). Similarly, Laura Shepherd and Laura Sjoberg (2012) argue that the neologisms 'cisprivilege' and 'cisgendered' are important for both rendering the social/cultural/political privileges of cis-identities visible and for combatting the simultaneous in/hypervisibility of trans-individuals.[17] Without drawing attention to, rendering visible, structures of privilege, restriction, dis/empowerment the ghostly traces go unattended to, the injustice continues. The consequences, the political implications of this simultaneous in/(hyper)visibility, therefore, is nothing less than erasure of the individuals' subjectivity. What Haunting does in the pursuit of ghosts is to engage with issues of in/(hyper)visibility *alongside* considerations of complex personhood, disturbed temporality, and power and it is this multi-faceted approach which enables us to better 'do justice' to the subjects of enquiry.[18]

Where the previous section has dealt with the way Haunting addresses the invisible as it occurs in traces, this section addresses the way in which ghosts can be made visible (as in known) through *feeling*. The ghosts here, rather than spectral shapes and shadows on the walls, are the kind of ghosts felt in the shiver-down-the-spine sensation, the sensation in looking at the familiar that something is different, or perhaps, something is missing.[19] Gordon is made uncomfortable by the *lack* of Sabina Spielrein in a photograph, saying 'a dead woman was not at a conference she was supposed to attend' drawing attention to her absence (2008, p. 42). Her absence becomes an uncomfortable presence that haunts Gordon, and forces her to keep seeking Spielrein's ghost and the ghosts that haunted Spielrein. Drawing on the theories of psychoanalysis that wove through Spielrein's life, Gordon uses 'the uncanny' to critique Freud's relationships with Jung and Spielrein, which then presents an inversion of Freud's uncomfortable experience of seeing 'an alienating figure' in the mirror that turns out to be his reflection.[20] Unlike the 'alienating figure' that is Freud's reflection, Spielrein emerges as a second reflection of Freud, an unacknowledged contributor to the development of a field that was instrumental in the articulation of her own mental illness. Through this ghost hunt, Gordon exposes the power at play in these lives through the patient–analyst relationship, the place of women in academia in the early development of psychoanalytic theory, and the (erroneous?) diagnosis of mental disorders specific to women.[21]

The project that formed the basis of this book grew from the same sense of discomfort. I wanted to find out what it was about drone warfare that made people so uncomfortable, what it was about crewing those machines that meant the individuals involved were viewed as particularly or peculiarly contemptible. I could see the flickers of something on the peripheries of my vision and as I looked into this problem, my 'gender!' alarm was beginning to squawk. Therefore Haraway's dictum of sitting with the trouble required that I both follow the traces of gender in the lives of British Reaper crews but also utilised Haunting to engage more fully with those traces. Gordon's discomfort led her to explore multiple (cob)webs of power at play in Spielrein's life: structural power in the form of whiteness and sexism, the biopower of medical discourses (as discussed at length by Michel Foucault), the label of 'madness', and the power of that 'madness' itself in enabling and inhibiting Spielrein's actions.[22] I wondered, as I started to look at the different accounts of Reaper crews, what webs of power I might discover, might feel, what sensuous knowledges I might be able deploy to investigate my discomfort.

Disturbed temporality

The complex persons that I trace are not just situated in a particular spatial and social situation, but also in a particular time. Using Haunting as a framework means rejecting the notion of linear temporal development and returning to Derrida's claims that '*Le temps est hors de ses gonds*' [time is off its hinges/is out of joint] (Derrida, 2006, p. 22) to illustrate the intermingling of the past, present and future. The 'present' becomes infused with the 'presence' of ghosts who require us to consider the multiplicity of presents: 'past present, actual present: "now", future present' in a manner Derrida refers to as the 'spectral moment' (Derrida, 2006, p. xix; see also Kenway, 2006; Coddington, 2011). As such, this past/present/future construction of the self includes interweavings of history, culture, myth, and fantasy.

The past does not just linger on in the present, but also can suggest alternative realities, the present other than as it is. The imagining of alternative 'presents' is part of how Gordon uses Haunting as a form of resistance against social injustice: The ghosts show us where the pain, power and dispossession are so that we can *act* to make changes.[23] The way that the past is remembered/memorialised as traumatic or edenic, the way that the future is perceived as promising or problematic are both things that interact with how the present is experienced now. As Derrida notes, ghosts may be historical, but they are 'not *dated* ... never docilely given a date in the chain of presents ...', their violence and our sense of discomfort comes from the disruption of our 'now' by the past, dislocating the progressive linearity of time (Derrida, 2006, p. 3).

To understand the importance of (the disturbance of) temporality, I deploy Peter Buse and Andrew Stott's (1999) articulation of anachronism. Beginning from the point that, in contemporary terms, any interest in the ghostly is considered anachronistic, they argue that this has *always* been the case and that this makes ghosts particularly useful for offering alternative readings of what has been, what is and what might be:

> ghosts are anachronism *par excellence*, the appearance of something in a time in which they clearly do not belong. But ghosts do not just represent reminders of the past ... they very often demand something in the future.
>
> (Buse and Stott, 1999, p. 14)

I found this description particularly useful as I was struck during the writing of this book by the appearance of the myth of the warrior. It felt nonsensical in the contemporary context of techno-warfare, of unmanned aerial vehicles, for the ghostly shapes of the historical/mythological figure of the warrior to linger in the sidelines of the debate. It is the presence of this figure, and the way it disturbs time and is riven through with particular gendered implications that I explore in the next chapter and weave through the following three chapters about British Reaper crews.

In this book, I note through all of the different data I use; be they interviews, memoirs, myths and legends, newspaper reports, cartoons or parliamentary debates, the myriad ways in which the past – who we *were* – is implicated in who we *are*. And the past is not always the historical past: but rather reality is woven through with fiction: myth is almost as (and sometimes more so) powerful than the histories. The future is viewed through eyes coloured by ideas of how the right way to be is predicated on the statement that *we've always done it this way*, as much as a hopefully smiling teleological endpoint. The ghost of the warrior who is traced through the following chapter emerges through centuries of storytelling about what it means to go to war, creating layers and layers of translucent mythology through which we view the current crop of military technological developments. Similarly, as much as history haunts the narratives of the now, much of the concern about military technology is laced with fear about its development in the future. The shuddering recourse to narratives about *The Terminator* and *The Matrix* reflect the way that we cannot stand still in the now, but rather must build relationships with our future selves before we have even truly imagined them.[24]

(Cob)webs of power

The final theme in my framework of Haunting is that of power. Power is woven through the other themes – the complexity of personhood, the way that intersectional markers dis/empower; the things we can/cannot/refuse to see and how what is seen is connected to what we (can) know; historical events and trends mark our pasts and the possibilities for the future, our realities and our fictions.[25] Power is then the last theme, but by no means the least. Gordon notes that

> the power relations that characterize any historically embedded society are never as transparently clear as the names we give to them imply. Power can be invisible, it can be fantastic, it can be dull and routine. It can be obvious, it can reach you by the baton of the police, it can speak the language of your thoughts and desires. It can feel like remote control, it can exhilarate like liberation, it can travel through time, and it can drown you in the present. It is dense and superficial, it can cause bodily injury, and it can harm you without seeming ever to touch you. It is systematic and it is particularistic and it is often both at the same time.
>
> (2008, p. 3)

An interest in the multiple shapes of power and illuminating the work that it does, in keeping with a poststructural standpoint, is at the centre of the framework of Haunting. Weaving together the concerns of complex personhood, in/(hyper) visibility, and temporality; Haunting foregrounds the way that power functions in big and small ways, at the structural and personal level, both material and ideational.

The importance of power and of tracing the web of its functions is a core concern to feminist scholars. Power is gendered, so that what is associated with the masculine is reified over the subordinated feminine. However, the traces of gendered power are not always (almost never) distinct and the way that gender functions to discipline and enable is traceable through the barely visible, the reference to (historically constructed) 'naturalness', the different ways in which masculinity and femininity are performed and understood

The final chapter of *Ghostly matters* traces a number of ghosts through Toni Morrison's (2005) *Beloved*. *Beloved* tells the story of a mother who kills her baby to prevent the baby from becoming a slave and how the return of the baby in the form of another woman causes ripples in families and local communities. This novel parallels the real-life case of Margaret Garner who, faced with being returned to a life of slavery, killed her daughter and tried to kill her sons, rather than see them go back to their owner. Beyond the haunting of the mother by a dead daughter, *Beloved*, and Gordon's analysis, is about 'the lingering inheritance of racial slavery' and the challenge it presents to ideas of progress (Gordon, 2008, p. 139). As such, the violence of the ghosts in this story are twofold: 'a slave mother's killing of her child and slavery' (Gordon, 2008, p. 141). Ghosts are then personally painful and painfully impersonal, they are subject and structure, power, pain and violence rolled into something that renders the familiar (a mother's love for her children) strange and uncomfortable (Freeman, 2010, p. 98). As such they may be a figure of an individual whose haunting represents a larger, structural ghost, or simply a ghost of that larger structure.[26] For example, Looking at the power of the capitalist economy, Derrida invites us to consider the 'apparition of the bodiless body of money: not the lifeless body or the cadaver, but a life without life ...' (Derrida, 2006, p. 51). This 'bodiless body' Derrida also refers to as 'spectralizing disincarnation' emphasising the fluid, mobile and 'becoming' nature of ghosts (Derrida, 2006, p. 51).

To summarise then, there are four key themes for my framework of Haunting: complex personhood, in/(hyper)visibility, disturbed temporality and power. These themes provide the focus for the methodology of the ghost hunt. Layered into and over one another, these themes inform the ghost hunt that I undertake in the following chapters as a means through which to investigate the interaction between gender and military technologies. Whilst the literature of the 'spectral (re)turn' indicates the plethora of sites to which Haunting can and has been applied, this has not, up to this point, included my own topic of interest. The subsequent section outlines in detail my rationale for choosing Haunting as a framework, followed by the ways in which I put the *methodology* of the ghost hunt to work.

Why Haunting for gendering military technologies?

At its core Haunting is about complexity: taking it seriously, trying as much as possible to embrace it; considering the implications of thinking of complexity as a central tenant of social science rather than an irritation in controlling variables. By so doing, it is possible to use the statement that 'life is complicated' to guide 'efforts to treat race, class, and gender dynamics and consciousness as more dense and delicate than those categorical terms often imply' (Gordon, 2008, p. 5). What the existing feminist literature on military technologies is currently missing is the specific ways in which this complexity challenges the way that gender binaries are currently thought about. Whilst poststructural feminists have clearly problematised the division between hierarchical masculine/ feminine dualisms, there remains a space for a framework that provides an alternative way of thinking, a way of thinking that can both acknowledge the important interjections of these scholars in critiquing the discourses of military masculinity, but also provide a way of understanding those uncomfortable facts that sit outside the 'either masculine or feminine' ways of thinking.[27] As I noted in Chapter 1, there is a fine line to be steered when thinking about the implications of technological developments in warfare between over and under-emphasising their effects.[28] There is also a wrestling match to be had over the extent to which a specific development, for example armed drones, constitutes something entirely novel, out of synch with its historic context and therefore, revolutionary; or whether the change is actually (technologically, politically, socially) marginal and being exaggerated for the sake of making bigger and more interesting academic/journalistic claims.[29] As noted in the introduction to this chapter, Haunting situates developments in military technology within their contextual particularity. It reveals, in ways that other frameworks cannot, that this particularity is laced with ghosts that disrupt time because the 'now' is always already haunted by histories and myths of the past, and by hopes and fears for the future.

The social context of warfare and the interaction between technologies and discourses of gender disrupt the 'common-sense' boundaries between life and death, between the 'real' and fictional, between the material and immaterial, between the present and the past.[30] As such, Haunting forces us to inhabit the liminal, shadowy spaces of impermanence that exist between the apparent oppositional dichotomies, what Zembylas (with reference to Papastephanou, 2011) eloquently describes as 'a peculiar "in-between" space that "reclaims the unspoken and neglected"' (2013, p. 79). Aligned to concerns about the 'unspoken and neglected' Haunting's focus on in/(hyper)visibility and presence/ absence as simultaneously occurring, and as implicated in (cob)webs of power, speaks directly to feminist theorising on the implications of these topics in relation to emancipation and empowerment from patriarchal gender structures.[31] As such, Ilya Parkins and Eva Karpinski argue that 'embracing complexity and contradictions in our own existence might help us to reckon with the spectre of feminist theory's own in/visibility in the present neoliberal moment ...' indicating the way in which the application of the framework of Haunting, with its focus on complexity (and indeed contradictions) can be useful to the development of thinking about gender and feminist theorising (2014, p. 6).

As a conceptual metaphor the ghost opens up 'complexity' through the possibility of 'ambivalent multiplicity – the reference to the liminal form of being (and thinking) encompassing life and death, human and non-human, presence and absence' (Blanco and Peeren, 2013, pp. 33, 91–92; see also Roberts, 2013, p. 8). This ambivalence suggests the possibility of logics beyond the binary, so that research is not

> reduced to a straightforward genesis, chronology or finitude [but rather] insists on blurring multiple borders, between visibility and invisibility, past and present, materiality and immateriality, science and pseudo-science, religion and superstition, life and death, presence and absence, reality and imagination ... challenging forms of authority.
>
> (Peeren, 2014, p. 10)

Undertaking a ghost hunt[32]

Ghost hunting is the methodology of Haunting through which the 'conceptual metaphor' of the ghost enables an exploration of complex personhood, in/(hyper) visibility, temporality and power. In examining these themes as they are approached through ghost hunting I introduce the concept of 'sensuous' knowledge, a core means through which the ghosts reveal themselves.

Janice Radway notes that previous approaches that focus on 'the visible and concrete' prevented adequate attention being paid to 'the particular density, delicacy, and propulsive force of the imagination' through which researchers can discover 'what has been lost' (Radway introducing Gordon, 2008, p. viii). For a framework committed to embracing the complexity of personhood in all its troubling, contradictory, and nuanced ways, attempting to rediscover 'what has been lost' through the exclusive use of scientific methods is a central concern. Therefore, the ontology of Haunting is more than the observable, requiring an engagement with what is *not* observable and indeed with what is imagined (Auchter, 2014; Peeren, 2014). Making reference to Raymond Williams' 'structures of feeling', that what is 'real' may also be invisible, sensual, troubling, and/ or 'seething' (Gordon, 2008, p. 19).

Engaging with 'sensuous' knowledges means taking seriously hunches, intuitions, unspoken commentary (like body language and expressions), popular portrayals of characters and identities, and artwork. Encapsulated under the title of 'sensuous knowledge', Gordon describes this kind of knowledge and interpretation as a

> mode of apprehension that notices and comprehends the ghostly matter of the sunken couch, the hat, the photograph, the reflection in the mirror, the open door ... a different kind of materialism, neither idealistic nor alienated,

but an active practice or passion for the lived reality of ghostly magical invented matters ... receptive, close, perceptual, embodied, incarnate.

(2008, p. 205)

Sensuous knowledge is important because of the way it affects our understanding of social reality because it 'always involves knowing and doing. Everything is in the experience with sensuous knowledge' (Gordon, 2008, p. 205). In embracing this framework and this methodology I was then forced to ask myself: How can I collect this kind of 'data'? What method(s) can I use?

What is considered 'knowledge' is important to 'ghost hunting' because the spectral, the sensation, the shiver have not previously been considered appropriate sources of information for social sciences. Undertaking a ghost hunt allows for engaging with the power inherent in discourses (regardless of whether they are fact or fiction), looking seriously at symbols, images, and photos, and listening to the echoes of the past and future in the troubling form of 'ghosts'.[33] Gordon discusses her answer to the question 'What method have you adopted for your research?', noting that in her case the question is really 'why do you use literary fiction as the "data" for your research and teaching and name this mode of knowledge production sociology, rather than, say, literary criticism?' (Gordon, 1990, p. 489). She thereby draws attention to the way that fact and fiction interact in the process of ghost hunting in ways that preclude the extraction of 'fact' and 'knowledge' as only those things that are externally real because 'our stories can be understood as fictions of the real' (Gordon, 2008, p. 11). That 'facts' are negotiated stories implies that multiple stories may be fact, that the process of their production involves power relations through which different stories may or may not be heard, and as a result it may not be possible to argue that any one story is more 'true' than any other.

Gordon uses a series of pieces of fiction as the discourses through which she searches for ghosts. In this book, I use interview data, journalist accounts, parliamentary debates, films and plays, and cartoons as my discourses. In keeping with the ghostly as unstable and fluctuating, I adopt Laura Shepherd's (2008a) approach to discourse, in as far as I view reality and discourse as interacting, creating, destroying, and engaged with one another. I do not view either 'reality' or 'discourse' as pre-existing the other, but rather as two components that dance together to create social reality and subjectivity. Shepherd's rejection of the 'distinction between a discursive and "non-discursive" realm' is particularly appropriate for ghost hunting which requires understanding the mobile meanings of discourses (2008, p. 18). Because the ghost is an unstable subject, the discourses, and the understanding of these discourses must be similarly mobile (Blanco, 2012). Ghost hunting then involves a close reading of various texts to see themes and motifs that are casting shadows from the past or that loom from the future. Taking Shepherd's (2008) view that reality is constructed out of discourses (as much as the reverse), ghost hunting views discourses as reconstructing the 'reality' of the past, rending it mythic, stripping out the things we no longer wish to see, but which remain as an uncomfortable presence. As such ghostly forms linger and disturb the narratives that we want to produce.

This project uses language as a site in which the shapes of the ghosts can be found. The discourses I use, the interview 'data', the transcripts of parliamentary debate, the cartoons are therefore the spaces in which I conjure up/engage with the ghostly shapes of the warrior, conversing about the work that gender is *doing* in these spaces. For example, asking 'how is the ghostly shape of the past, present and future tangled in these interpretations of gender?' 'How does this dis/empower individuals and groups?' 'How does it affect how novel military technologies are understood?' These are not simple questions with straight forward answers, but rather engaging with discourse in this manner allows the ghostly interlocuter to point to spaces on confusion, complication and contradiction. As Wendy Brown neatly summarises, the 'truth' that we inherit is

> not 'what really happened' to the dead but what lives on from that happening, what is conjured from it, how past generations and events occupy the force fields of the present, how they claim us, and how they haunt, plague, and inspirit our imaginations and visions for the future.
>
> (2001, p. 150)

The aim then, is to explore how and why specific iterations of the past live on in preference to others and to question the capacity of the living to speak for the dead.

In keeping with the poststructural feminism I adopt for this book, the 'posthumanist subject' of the ghost opens the gendered binaries of masculinities and femininities to critique in ways that are in keeping with the ethical commitments of both Haunting as a framework and feminism as a political project.[34] Whilst Haunting offers a means of destabilising, or gesturing to the instability of the distinction between the categories of gendered binaries, acknowledging that 'haunted spaces ... emerge through a specific refusal of classification' it currently lacks an adequate vocabulary to make sense of this opening (Holloway and Kneale, 2008, p. 308). In order to engage with this opening I suggest the addition of an interjection from queer theory. Specifically, the inclusion of Cynthia Weber's (2014b, 2016) 'queer logic(s)', as the vocabulary through which the potential for binary resistance in Haunting can be more adequately realised.

Queer theory and queer logic(s)

Queer theory grew out of the work of Evie Sedgwick (2008, 1993, 2000), Teresa de Lauretis (1986, 1987, 1990, 1991) and Judith Butler (1999, 2004b, 2011) and is now applied to a wide range of diverse subject areas: sexualities and identity politics, literature, legal theory, human resource development, international political economy, geography, and the area of this book: subjects relating to international security and war.[35] Like feminism, queer theory also critically engages with the power of binaries, reflecting on normal/perverse, particularly in the form of heterosexual/homosexual (Weber, 2016, p. 199). For example, Evie Sedgwick's list of binary distinctions relating to international security includes 'public/private, domestic/foreign, discipline/terrorism, secrecy/disclosure, natural/ artificial, wholeness/decadence and knowledge/ignorance' (Sedgwick in Weber, 2014, p. 598). Both queer theorists and post-structural feminists share the similar aim in highlighting such binaries which is not to reproduce them, but to reveal the instabilities, contradictions, and inherent 'queerness' within them. As such '[q]ueer IR scholars … track when queer figurations emerge and how they are normalized/perverted' with the aim of producing 'deviant knowledges' and queer logic(s) (Ahmed, 2006).

It is not possible to do justice to the entirety (or even a substantial portion) of the vast array of queer theory literature here, and the book aims to utilise only one component of queer thinking, which is Weber's (2016, 2014) 'queer logic(s)'.[36] Based on Roland Barthes (1974, 1977) 'plural logic', the idea of queer logic(s) is specifically related to challenging the kinds of problems with binaries noted in Chapter 1, with queer logic(s) providing a means of being and/ or rather than either/or. This means 'that one can be a boy or a girl while at the same time being a boy and a girl' (Weber, 2014, p. 598), and therefore 'queer subjectivities more than exceed binary logics of the either/or' (Weber, 2016, p. 166). Weber adds to Barthes logic by pluralising the very plural logic that he proposes, creating a '(pluralised) and/or' to exceed and queer the binary categories that Barthes invites us to challenge (Weber, 2016, Kindle location 909). This, Weber argues, allows for identities to be more than 'and' and/or 'or'. Approaching the concerns of International Relations in this way 'directs us … to categories that connect and break apart foundational binaries … by understanding the stabilising "slash"[37] in these binaries as multiplying and complicating connections, figures, and orders rather than reducing and stabilizing them' (Weber, 2016, Kindle location 948). As such, queer logic(s) troubles the binaries that establish as separate the categories of male and female, black and white, abled and disabled, etc.[38] Queer theorists shed light on the tenuousness of links that are considered concrete and rigid, asking why structures are made in the ways that they are, asking who benefits and who suffers, situating the categories in the social context in which they function, rather than arguing that they represent immutable and eternal truths.[39]

I include Weber's 'queer logic(s)' in my use of Haunting because, being neither dead nor alive, being present but also absent, ghosts already inherently embody this logic. As such, I want to render this connection explicit. As Derrida notes, ghosts are 'more than one/no more one [*le plus d'un*]', bringing '[an] either/or logic into question' (2006, p. xx). I am therefore not *making* Haunting reflect a queer logic(s): this logic was already built in to it. Rather I am choosing to highlight the queer logic(s) because it helps to draw out the ways in which the arguments of masculinity and femininity in relation to military technologies are limited because the experiences of the crews are always, already, 'exceeding all binary opposites' (Harris, 2015, p. 17). For those individuals who view the introduction of the drone as heralding a new age, or a revolution in military affairs, the ghosts serve to demonstrate the way 'the becoming-future is haunting us' through fears of terminators, swarms, and lethal autonomous robots (Puar, 2007). For individuals who are troubled by the way that drones captured the public zeitgeist and the way the arguments slid past each other so much like oil and water, the ghosts draw attention to how 'being haunted draws us affectively, sometimes against our will and always a bit magically' (Puar, 2007). The ghosts create a liminal shadowy space in which the crews can simultaneously be distant and present, be powerful and be impotent, be warriors and not be warriors. By drawing on the 'sensuous knowledge', and the 'peripheral things, sniping from the side-lines and the depths', I have tried in the following chapters to paint the complicated picture of the interaction between gender and military technologies in a way that rejects the feminised/masculinised dichotomy and embraces the queer logics of the ghost (Gordon, 2008, pp. 60, 205; Frosh, 2013).

Conclusion

This chapter has outlined the framework and methodology that I use in this book. Going back to the title of the introduction ('life's complicated') is useful because at its heart this is what Haunting is all about: complications, complexity, and contradictions. In this chapter, and throughout the rest of the book, I argue for the use of Haunting to research military technologies and gender because it provides a means through which to understand the data that does not 'fit' with conventional gender narratives. By using Haunting, as outlined in this chapter, it is possible to construct an assessment of the interaction(s) between gender and military technology in situated, particular, nuanced ways: 'connect[ing] ideas and events that at first appear unrelated' (Coddington, 2011, p. 744). In addition, by interrogating gender and military technology through the prism of Haunting I aim to illuminate how the individuals who engage with military technologies are always already haunted by ghosts of history, future, gender, intersectionality, instability; by ghosts that are here and there, absent and present, alive and dead.

The framework of Haunting, outlined in this chapter, emerges from the 'spectral (re)turn'. I have framed my understanding around the four core themes of complex personhood, in/(hyper)visibility, disturbed temporality, and power. Through these themes, Haunting provides an important means of engaging with the effects of intersectional identifiers and issues of interest to International Relations. This is particularly useful for studying the interaction between gender and military technologies because the introduction of drones invites the consideration of their position along so many binary distinctions: near/far, intimate/distant, historical fact/(science)fiction; and because the data about drones and their use appears to include elements that straddle the 'slash' between the two. Given Haunting's engagement with ghosts (figures who breech the dead/alive and present/absent boundaries) it provides a useful means of getting to the liminal, shadowy spaces where gender is doing work in this context.

The ghost hunt is the 'method' of Haunting, a way of paying attention to sensuous knowledges those sources of information outside of the traditional 'visible and concrete', in order to engage the ghosts in 'conversation' as a means of gaining additional insights in to the work that gender does (Radway cited in Gordon, 2008, p. viii). 'Sensuous knowledge' then, is the way in which ghost hunting embraces complexity with the aim of destabilising binaries, however it is lacking the necessary vocabulary through which to make this explicit. As a result, I have added to the ghost hunt the vocabulary and theorising of queer logic(s). The inclusion of Weber's queer logic(s) is the important final component required to make the ghost hunt a useful methodology for interrogating the interaction between gender and military technologies. Rendering explicit the ways in which intersectional identifiers can, and do, exceed *either/or* distinctions, the inclusion of queer logic(s) provides the language to explore logics of *and/or* which emerge through discussions of the work that gender does in the lives of British Reaper crews. These crews are situated against a specific contextual backdrop of history, culture and myth; and in the next chapter I explore this backdrop through a ghost hunt of the figure of the warrior – the masculine ghost of the histories, myths and legends of war through the ages and across space.

Notes

1   (Krista Benson quoted in Marinucci, 2010, p. xv).
2   As noted in the introduction, I capitalise Haunting when I am referring to the theoretical framework I sketch this chapter and use 'haunting' when using the term otherwise.
3   One of the things that makes the patriarchy and sexism so powerful is their 'stickiness', they are difficult to overcome, they get stuck all over you, you can become trapped in them – the more you try to unravel and remove yourself from their webs the more tangled you become. (For discussions of gender/sexuality as 'sticky' see Puar, 2007; Ahmed, 2006; Zalewski and Runyan, 2013).
4   Luckhurst speaks of the spectral turn, but in keeping with Derrida's statement that the spectre is always re-turning I have amended the language to reflect the specific nature of the ghost.
5   'Stuff' is the term that Laura Shepherd (2013) uses to describe her 'data', and I re-use it here because 'data' is too concrete a word for describing the ephemeral traces of ghosts.
6   There are various different ways in which Haunting is understood or used, I will explore some of these in the subsequent sections of the chapter.
7   As noted in the previous chapter, femininity is the important binary opposite to masculinity. However, because masculinity is the most important marker in the British military it is masculinity that forms the focus of this book.
8   See section on Complex Personhood in this chapter.
9   (Bergmann, 2011; Georgescu, 2011; Razinsky, 2010).
10   In language that reflects Freud's concept of the uncanny, Auchter notes 'The spectre is familiar to us precisely because it is a remnant of ontology, of the creation of being.... Haunting, then, de-ontologises' (2014, p. 17).
11   (Brown, 2001; Gordon, 2008; Gunn, 2006; Szeman, 2000).
12   Gordon is interested by Williams' search for the history of her great-great grandmother who was a slave. This is a history that Williams traces by looking for the shape of her absence in the reports and writings of her owner.

13 I am indebted to Rhiannon Neilsen for suggesting this organisational strategy that has become (!) so useful and central to my understanding of 'complex personhood'.

14 Written notes from Rhiannon Neilsen on an earlier iteration of this section.

15 Using 'known' here is somewhat problematic as it appears to imply a certain concreteness, a wholeness. Using the term is, as Haraway (2017) notes, a non-innocent pursuit, therefore, although I use 'known' here please note that I use it as a shorthand for the things which whilst only partially graspable are nonetheless the types of 'data' that this project pursues.

16 (Enloe, 2013; Sjoberg, 2007; Wilcox, 2014).

17 Shepherd and Sjoberg usefully describe cisprivilege as 'the privilege enjoyed by people who identify wholly with, feel comfortable in, are seen to belong to or "are" the gender/sex they are assigned at birth and/or raised to believe that they "are"' (2012, p. 6).

18 Both human subject and subject as topic of study.

19 (Holloway and Kneale, 2008; Parkins and Karpinski, 2014; Engle, 2009).

20 (Gordon, 2008, p. 54; see also Certeau, 1986; Freccero, 2006).

21 (Gordon, 2008; Appignanesi, 2009; Kaplan, 2011; Frosh, 2013).

22 (Foucault, 1991; Foucault and Howard, 1971; Foucault and Senellart, 2010).

23 (Gordon, 2008; see also Coddington, 2011; Saleh-Hanna, 2015).

24 (Holmes, 2015; Turse and Engelhardt, 2012).

25 (Coddington, 2011; Cruz and Frijhoff, 2009; Appignanesi, 2009; Auchter, 2013).

26 For those of you who enjoy Terry Pratchett, it might be instructive to consider Granny Weatherwax's discovery of the 'mind of the Kingdom' in *Wyrd Sisters*.

27 (for example, Peterson and Runyan, 2010; Shepherd, 2009; Sjoberg, 2014; Sylvester, 1994).

28 (Echevarria, 2007; Hundley, 1999).

29 (Lee, 2013; Dunn, 2013; Enemark, 2015).

30 (Brown, 2001; Freccero, 2006; Gordon, 2008).

31 (Bridenthal *et al.*, 1998; Hesford, 2011; Keller, 1986).

32 I feel a little uncomfortable using the term ghost hunt because of the implications of the term 'hunt'. The aim of a hunt is often the death or eradication of sometime, an animal perhaps, and it implies a certain level of violence. However, I am using the term here to mean a hunt in the sense of seeking (hunting for my lost keys), and continue to aim to meet the ghosts I find with hospitality rather than an aim to destroy.

33 (Kronsell *et al.*, 2006; Gordon, 2008).

34 (Gunn, 2006, p. 83; Gordon, 2008; Auchter, 2014; Blanco and Peeren, 2013; Peterson and Runyan, 2010).

35 (for example Carlin, 2012; Fineman *et al.*, 2009; Gedro and Mizzi, 2014; Griffin, 2009; Smith, 2012; Binnie, 2009; Shepherd and Sjoberg, 2012; Weber, 2014; Wilcox, 2014).

36 An excellent summary is available (Richter-Montpetit and Weber, 2017). It is somewhat contentious to use queer theory in project that is not explicitly linked to nonnormative, or queer, sexualities and the debate regarding this can be found in (Richter-Montpetit, 2018).

37 Weber uses the term 'slash' in her description of queer logic(s). It is a slightly strange term but one which I think usefully illustrates the thinnest of the distinction between the two categories and also gestures towards the violence which can result from disciplining subjects to fit into one of these boxes.

38 (Wiegman, 2006; Halberstram, 2007; Jackson, 2006; Browne, 2012; Munoz and Browne, 2012).

39 (Browne, 2012; Puar, 2007; Peterson and Runyan, 2010).

References

Abraham, N. and Torok, M. (1999) *Cryptonymie: le verbier de l'homme aux loups*. Champs 425. Paris: Flammarion.

Abraham, N., Torok, M., and Rand, N. (2008) *L'écorce et le noyau*. Paris: Flammarion.

Adorno, T.W. (1974) Theses against occultism. *Telos*, 1974 (19): 7–12. doi: 10.3817/0374019007.

Agamben, G. (1998) *Sovereign power and bare life*. Homo sacer 1. Stanford, CA: Stanford University Press.

Ahmed, S. (2006) *Queer phenomenology*. Durham, NC: Duke University Press.

Appignanesi, L. (2009) *Mad, bad and sad: a history of women and the mind doctors from 1800 to the present*. London: Virago.

Auchter, J. (2013) Border monuments: memory, counter-memory, and (b)ordering practices along the US–Mexico border. *Review of International Studies*, 39 (02): 291–311.

Auchter, J. (2014) *The politics of haunting and memory in international relations*. Interventions. London; New York: Routledge/Taylor & Francis Group.

Bergmann, M.S. (2011) The dual impact of Freud's death and Freud's death instinct theory on the history of psychoanalysis. *The Psychoanalytic Review*, 98 (5): 665–686. doi: 10.1521/prev.2011.98.6.665.

Binnie, J. (2009) Epistemology, methodology, and pedagogy. In Browne, K., Lim, J. and Brown, G. (eds) *Geographies of sexualities: theory, practices and politics*. Ashgate. pp. 29–38.

Blanco, M. del P. (2012) *Ghost-watching American modernity: haunting, landscape, and the hemispheric imagination*. 1st edn. New York: Fordham University Press.

Blanco, M. del P. and Peeren, E. (eds) (2013) *The spectralities reader: ghosts and haunting in contemporary cultural theory*. New York: Bloomsbury Academic.

Braidotti, R. (2002) *Metamorphoses: towards a materialist theory of becoming*. Cambridge, UK; Malden, MA: Published by Polity Press in association with Blackwell Publishers.

Bridenthal, R., Stuard, S. and Wiesner-Hanks, M.E. (1998) *Becoming visible: women in European history*. 3rd edn. Andover, UK: Cengage.

Brown, Wendy (2001) *Politics out of history*. Princeton, NJ: Princeton University Press.

Browne, K. (ed.) (2012) *Queer methods and methodologies: intersecting queer theories and social science research*. Farnham, Surrey: Ashgate.

Buse, P. and Stott, A. (1999) *Ghosts: deconstruction, psychoanalysis, history*. New York: St Martin's Press. Available at: http://site.ebrary.com/id/10612853 (accessed: 29 January 2018).

Butler, J. (1999) *Gender trouble feminism and the subversion of identity*. New York: Routledge. Available at: http://kcl.etailer.dpsl.net/home/html/moreinfo.asp?isbn=0203902750&whichpage=1&pagename=category.asp (accessed: 17 September 2014).

Butler, J. (2004a) *Precarious life: the powers of mourning and violence*. London; New York: Verso.

Butler, J. (2004b) *Undoing gender*. New York; London: Routledge.

Butler, J. (2011) *Bodies that matter: on the discursive limits of 'sex'*. Routledge classics. Abingdon, Oxon; New York: Routledge.

Carlin, D. (2012) Queer Theory and the American novel. In Bendixen, A. (ed.) *A companion to the American novel* [online]. Chichester, UK: John Wiley & Sons, Ltd. pp. 342–356. Available at: http://doi.wiley.com/10.1002/9781118384329.ch20 (accessed: 22 November 2015).

Certeau, Michel de (1986) *Heterologies: discourse on the other*. Minneapolis: University of Minneapolis Press.

Coddington, K.S. (2011) Spectral geographies: haunting and everyday state practices in colonial and present-day Alaska. *Social & Cultural Geography*, 12 (7): 743–756. doi: 10.1080/14649365.2011.609411.

Comaroff, J. (2007) Ghostly topographies: landscape and biopower in modern Singapore. *Cultural Geographies*, 14 (1): 56–73. doi: 10.1177/1474474007072819.

Cosgrove, D.E. (2008) *Geography and vision: seeing, imagining and representing the world*. International library of human geography v. 12. New York : New York: Palgrave Macmillan.

Cruz, L. and Frijhoff, W. (eds) (2009) *Myth in history, history in myth. Brill's studies in intellectual history v. 182*. Leiden; Boston: Brill.

Davis, C. (2005) Hauntology, spectres and phantoms. *French Studies*, 59 (3): 373–9.

Derrida, J. (2006) *Specters of Marx: the state of the debt, the work of mourning and the New International*. Routledge classics. 1. New York: Routledge.

Dunn, D.H. (2013) Drones: disembodied aerial warfare and the unarticulated threat. *International Affairs*, 89 (5): 1237–1246.

Echevarria, A.J. (2007) *Imagining future war: the West's technological revolution and visions of wars to come, 1880–1914. War, technology, and history*. Westport, CT: Praeger Security International.

Edensor, T. (2005) *Industrial ruins: spaces, aesthetics, and materiality*. Oxford, UK; New York: Berg.

Enermark, Christian (2015) Armed drones and the ethics of war: military virtue in a post-heroic age. Routledge.

Enloe, Cynthia (2013) *Seriously! Investigating crashes and crises as if women mattered*. Berkeley, CA: University of California Press.

Fineman, M., Jackson, J.E. and Romero, A.P. (eds) (2009) *Feminist and queer legal theory: intimate encounters, uncomfortable conversations*. Farnham, Surrey, England; Burlington, VT: Ashgate.

Foucault, M. (1991) *Discipline and punish: the birth of the prison*. Penguin social sciences. Reprint. London: Penguin Books.

Foucault, M. and Howard, R. (1971) *Madness and civilization: a history of insanity in the Age of Reason*. Social science paperback 72. London: Tavistock Publications.

Foucault, M. and Senellart, M. (2010) *The birth of biopolitics: lectures at the Collège de France, 1978–79*. Michel Foucault's lectures at the Collège de France. Basingstoke: Palgrave Macmillan.

Freccero, C. (2006) *Queer/early/modern*. Series Q. Durham: Duke University Press.

Gunn, J. (2006) Review essay: mourning humanism, or, the idiom of haunting. *Quarterly Journal of Speech*, 92 (1): 77–102.

Freeman, Elizabeth (2010) *Time binds: queer temporalities, queer histories*. Durham: Duke University Press.

Freud, S. (1997) The uncanny. In *Writings on art and literature*. Stanford, CA: Stanford University Press. pp. 193–233.

Frosh, S. (2013) *Hauntings: psychoanalysis and ghostly transmissions*. Houndmills, Basingstoke, Hampshire; New York: Palgrave Macmillan.

Garrett, B.L. (2011) Videographic geographies: using digital video for geographic research. *Progress in Human Geography*, 35 (4): 521–541. doi: 10.1177/0309132510388337.

Gedro, J. and Mizzi, R.C. (2014) Feminist theory and queer theory: implications for HRD research and practice. *Advances in Developing Human Resources*, 16 (4): 445–456.

Georgescu, M. (2011) Freud's theory of the death drive. *Review of Contemporary Philosophy*, 10: 228–233.

Gordon, A. (1990) Feminism, writing and ghosts. *Social Problems*, 37 (4): 485–500.

Gordon, A. (2008) *Ghostly matters: haunting and the sociological imagination*. New University of Minnesota Press ed. Minneapolis, MN: University of Minnesota Press.

Griffin, P. (2009) *Gendering the World Bank: neoliberalism and the gendered foundations of global governance*. Basingstoke, UK; New York: Palgrave Macmillan.

Halberstram, J. (2007) Transgender butch: butch/FTM border wars and the masculine continuum. *GLQ: A Journal of Lesbian and Gay Studies*, 4 (2): 287.

Haraway, D.J. (2017) *Staying with the trouble making kin in the chthulucene*. Old Saybrook, CT: Tantor Media. Available at: http://rbdigital.rbdigital.com (accessed: 21 December 2017).

Harris, V. (2015) Hauntology, archivy and banditry: an engagement with Derrida and Zapiro. *Critical Arts*, 29 (sup1): 13–27.

Hawkins, H. (2010) 'The argument of the eye'? The cultural geographies of installation art. *Cultural Geographies*, 17 (3): 321–340. doi: 10.1177/1474474010368605.

Hesford, W. (2011) *Spectacular rhetorics: human rights visions, recognitions, feminisms*. Durham, NC: Duke University Press.

Holloway, J. and Kneale, J. (2008) Locating haunting: a ghost-hunter's guide. *Cultural Geographies*, 15 (3): 297–312. doi: 10.1177/1474474008091329.

Holmes, K. (2015) Is today's drone tomorrow's Skynet, Terminator? *CNS News* [online], 5 May. Available at: www.cnsnews.com/commentary/kim-holmes/todays-drone-tomorrows-skynet-terminator (accessed: 3 July 2016).

Jackson, S. (2006) Heterosexuality, sexuality and gender. In Richardson, D., McLaughlin, J. and Casey, M.E. (eds) *Intersections between feminist and queer theory*. London: Palgrave Macmillan. pp. 38–58.

Jacob, C. and Dahl, E.H. (2006) *The sovereign map: theoretical approaches in cartography throughout history*. English-language ed. Chicago, IL: University of Chicago Press.

Kaplan, R.M. (2011) Mad, bad and sad: a history of women and the mind doctors from 1800 to the present. Appignanesi, L. (2009). Book review. *Australasian Psychiatry*, 19 (2): 173–174.

Keller, E.F. (1986) Making gender visible in the pursuit of nature's secrets. In de Lauretis, T. (ed.) *Feminist studies/critical studies*. Bloomington, IN: Indiana University Press. pp. 67–77.

Kenway, J. (2006) *Haunting the knowledge economy*. London; New York: Routledge.

Kleinberg, E. (2017) *Haunting history: for a deconstructive approach to the past*. Meridian: crossing aesthetics. Stanford, CA: Stanford University Press.

Kronsell, A (2006) Methods for studying silences: gender analysis in institutions of hegemonic masculinity. In Ackerly, B.A. and Stern, M. (eds) *Feminist methodologies for international relations*. Cambridge: Cambridge University Press. pp. 108–128.

Lamble, S. (2009) Unknowable bodies, unthinkable sexualities: lesbian and transgender legal invisibility in the Toronto women's bathhouse raid. *Social & Legal Studies*, 18 (1): 111–130. doi: 10.1177/0964663908100336.

de Lauretis, T. (1986) Feminist studies/critical studies: issues, terms and contexts. In de Lauretis, T. (ed.) *Feminist studies/critical studies*. London: Macmillan. pp. 1–19.

de Lauretis, T. (1987) *Technologies of gender: essays on theory, film and fiction*. London: Macmillan.

de Lauretis, T. (1990) Eccentric subjects: feminist theory and historical consciousness. *Feminist Studies*, 16 (1): 115–150.

de Lauretis, T. (1991) Queer theory: lesbian and gay sexualities. *Differences*, 3 (2): iii–xvii.

Lee, P. (2013) Rights, wrongs and drones: remote warfare, ethics and the challenge of just war reasoning. *Air Power Review*, 16 (3): 30–49.

Luckhurst, R. (2002) The contemporary London gothic and the limits of the 'spectral turn'. *Textual Practice*, 16 (3): 527–536.

Marinucci, M. (2010) *Feminism is queer: the intimate connection between queer and feminist theory*. London: Zed.

Mayerfield Bell, M. (1997) The ghosts of place. *Theory and Society*, 26 (6): 813–836.

Morrison, Toni (2005) *Beloved*. London: Vintage,.

Munoz, L. and Browne, K. (2012) Brown, queer and gendered: queering the Latina/o 'street-scapes' in Los Angeles. In *Queer methods and methodologies: intersecting queer theories and social science research*. Surrey: Ashgate. pp. 55–68.

Papastephanou, Marianna (2011) Material specters: international conflicts, disaster management, and educational projects. *Educational Theory*, 61 (1): 97–115.

Peeren, E. (2014) *The spectral metaphor*. London: Palgrave Macmillan. Available at: www.palgraveconnect.com/doifinder/10.1057/9781137375858 (accessed: 28 April 2016).

Peterson, V.S. and Runyan, A.S. (2010) *Global gender issues in the new millennium. Dilemmas in world politics*. 3rd edn. Boulder, CO: Westview Press.

Puar, J.K. (2007) *Terrorist assemblages: homonationalism in queer times*. Next wave. Durham: Duke University Press.

Razinsky, L. (2010) Driving death away: death and Freud's theory of the death drive. *The Psychoanalytic Review*, 97 (3): 393–424. doi: 10.1521/prev.2010.97.3.393.

Richter-Montpetit, M. (2018) Everything you always wanted to know about sex (in IR) but were afraid to ask: the 'queer turn' in international relations. *Millennium: Journal of International Studies*, 46 (2): 220–240. doi: 10.1177/0305829817733131.

Richter-Montpetit, M. and Weber, C. (2017) *Queer international relations*. Oxford: Oxford University Press. doi: 10.1093/acrefore/9780190228637.013.265.

Roberts, E. (2013) Geography and the visual image: a hauntological approach. *Progress in Human Geography*, 37 (3): 386–402. doi: 10.1177/0309132512460902.

Roseneil, S. (2009) Haunting in an age of individualization: subjectivity, relationality and the traces of the lives of others. *European Societies*, 11 (3): 411–430. doi: 10.1080/14616690902764823.

Salih, S. (2002) *Judith Butler*. Routledge critical thinkers. London; New York: Routledge.

Sedgwick, E. (1993) *Tendencies*. Durham, NC: Duke University Press.

Sedgwick, E. (2000) *A dialogue on love*. Boston, MA: Beacon Press.

Sedgwick, E. (2008) *Epistemology of the closet*. Berkeley, CA: University of California Press.

Shepherd, L.J. (2008) *Gender, violence and security: discourse as practice*. London; New York: Zed Books; Distributed exclusively in the USA by Palgrave Macmillan.

Shepherd, L.J. (2009) Gender, violence and global politics: contemporary debates in feminist security studies. *Political Studies Review*, 7 (2): 208–219.

Shepherd, L.J. (ed.) (2013) *Critical approaches to security: an introduction to theories and methods*. London; New York: Routledge.

Shepherd, L.J. and Sjoberg, L. (2012) Trans-bodies in/of war(s): cisprivilege and contemporary security strategy. *Feminist Review*, 101: 5–23.

Simpson, P. (2011) 'So, as you can see …': some reflections on the utility of video methodologies in the study of embodied practices: video methodologies in the study of embodied practices. *Area*, 43 (3): 343–352. doi: 10.1111/j.1475-4762.2011.00998.x.

Sjoberg, L. (2007) *Mothers, monsters, whores: women's violence in global politics*. New York: Zed Books.

Sjoberg, L. (2010) Gendering the Empire's soldiers: gender ideologies, the U.S. military, and the 'War on Terror'. In Sjoberg, L., Via, S. and Enloe, C. (eds) *Gender, war and militarism: feminist perspectives*. Santa Barbara, CA: Praeger.

Szeman, I. (2000) Ghostly matters: On Derrida's *specters. Rethinking Marxism*, 12 (2): 104–116.

Tolia-Kelly, D. (2004) Locating processes of identification: studying the precipitates of re-memory through artefacts in the British Asian home. *Transactions of the Institute of British Geographers*, 29 (3): 314–329. doi: 10.1111/j.0020-2754.2004.00303.x.

Turse, N. and Engelhardt, T. (2012) *Terminator planet: the first history of drone warfare, 2001–2050*. Lexington, KY: Dispatch Books.

Weber, C. (2014) From queer to queer IR. *International Studies Review*, 16 (4): 596–601. doi: 10.1111/misr.12160.

Weber, C. (2016) *Queer international relations: sovereignty, sexuality and the will to knowledge*. Oxford studies in gender and international relations. New York: Oxford University Press.

Welland, J. (2013) Militarised violences, basic training, and the myths of asexuality and discipline. *Review of International Studies*, 39 (04): 881–902. doi: 10.1017/S0260210512000605.

Wilcox, L. (2014) Queer theory and the 'proper objects' of international relations. *International Studies Review*, 16 (4): 612–615.

Zalewski, M. (2000) *Feminism after postmodernism: theorising through practice*. London; New York: Routledge.

Zalewski, M. and Runyan, A.S. (2013) Taking feminist violence seriously in feminist international relations. *International Feminist Journal of Politics*, 15 (3): 293–313. doi: 10.1080/14616742.2013.766102.

Zembylas, Michalinos (2013) Pedagogies of hauntology in history education: learning to live with the ghosts of disappeared victims of war and dictatorship: pedagogies of hauntology in history education. *Education Theory*, 63 (1): 69–86.

3   H(a)unting the warrior

Introduction

Concerns about the 'feminisation' of contemporary Western militaries are often framed in terms of the loss of the warrior spirit.[1] This 'warrior spirit', the commentators argue, is what divides soldiers (who fight in pursuit of a wider, socially desirable cause) from mercenaries (who fight instrumentally for money) (Braudy, 2005). Warrior spirit is also what divides the individual who is a soldier because he is conscripted and the individual who is a soldier because it enables him to become a better version of himself.[2] In these narratives about the warrior, the relationship between the individual and military technology is expressed in the relationship between the warrior and his weapon – be that Excalibur or the 'Fat Man' atomic bomb. The changeable nature of the ghost of the warrior as part of this interaction with technology is, for example, illustrated by Gardiner who stated 'The warrior identity itself is ... an inherently unstable construction. As a form of masculinity, it demands constant testing. The results are predictably contested' (Gardiner, 2012, p. 380) and by Robert L. O'Connell who noted:

> ... the change in the ideal warrior personality is wrought by the advent of guns in the fifteenth and sixteenth centuries, from 'ferocious aggressiveness' to 'passive disdain'. So there is no personality type – 'hot tempered', 'macho', or whatever – consistently and universally associated with warfare.
>
> (Cited in Ehrenreich, 1998, p. 9)

The type of military technology, therefore, is implicated in the kind of behaviour and attributes deemed desirable in the warrior, and different kinds of masculinity which needed to be performed to gain the warrior status.

Before I begin, it is necessary to add the following caveat: this chapter lays out a gendered reading of various iterations of the warrior as masculine (and heteronormative) as though this were unproblematic. This is not the case, but I do so in order to illustrate the 'traditional' (not necessarily historically accurate) backdrop against which I situate the lives and experiences of British Reaper crews. This is because these 'traditional' iterations have become the dominant

narratives. In the empirical chapters of this book I will begin to unpick and prob-
lematise some of the binaries and gendered 'stories' about the warrior. Therefore, the purpose of this chapter is to lay out a reading that I will use my
empirical chapters to critique.

With this in mind, the chapter argues that the ghost of the warrior acts as a
'semiotic [trope] that combine[s] knowledges, practices, and power to shape how
we map our worlds and understand actual things in those worlds' (Weber, 2016,
Kindle location 663). In so doing, the chapter sets out some different ways and
places that the ghost of the warrior appears, illuminating the way that the current
styles and structures of military masculinity are laid across historical/mythologi-
cal accounts, figures from the past/possibilities for the future, social/cultural/
psychological elements and hierarchies that dis/empower. Barbara Ehrenreich
provides some illustrations which demonstrate the pertinence of the ghostly to
the study of the warrior, claiming

> It is from [a] notion of lineage that the warrior derived his most exalted and
> mystic sense of who he was: not merely a mortal individual, 'born of
> woman', but a link within a far superior tradition, analogous to a priesthood
> and composed exclusively of men, meaning often only 'noble' men.
>
> (1998, p. 151)

And it is noted that General Patton 'believed he was the reincarnation of dead
heroes, both confederate and Viking' and that 'Nineteenth-century Prussians …
saw themselves as the successors of the Spartan and the Assyrians, or rooted
their pedigree in medieval times, sometimes describing themselves as crusaders'
(Ehrenreich, 1998, pp. 152, 153). The haunting of the contemporary military by
the ghosts of 'warriors past' is woven into the very fabric of the institutions. For
example, as Victoria Basham notes of her fieldwork 'The dining hall, where I
had lunch, is full of long wooden tables and painted portraits of long dead
military men' (2013, p. 1). As a result the literature on 'the warrior' ranges
across ancient history, anthropology, sociology, and military culture and the
ghost of the warrior appears and reappears throughout ancient and modern literature, art, political accounts, and poetry.[3] Therefore, research into the warrior
reveals that there is a plethora of 'quintessential' warriors, and as such, there is
no *one* original warrior but a multitude.[4]

As the Gardiner quote at the start of the chapter illustrates, the figure of the
warrior is 'inherently unstable', being constantly constructed, deconstructed and
reconstructed and therefore the ghostly warriors outlined in this chapter are not
an exhaustive collection but function as an 'amuse bouche' of possible iterations.
The space constraints mean that each sketch is necessarily limited and incomplete, but represents an attempt to give a flavour of the differences and similarities. This chapter traces the spectral and shadowy stories of different warrior
masculinities through the following iterations: the Classical Greek, the Brother-
at-Arms, the Chivalrous Gentleman, the Warrior as Lover, the Beautiful Hero
and the Husband–Protector. I have chosen these because they cover a range of

historical periods, geographical, cultural, and social contexts.[5] These warriors were also chosen because in their ghostliness they continue to have an impact on the understanding of what it means to be a warrior and to enact military masculinity today. As such, the chapter concludes with a reflection on the overarching theme of masculinity that interweaves through the various figurations outlined in the ghost hunt. This theme is an implicit component of the warrior figure in all its iterations, but the role of gender and the sexed/gendered body is rarely explicitly acknowledged within the literature. The ghost hunt then functions to illustrate the patchwork heritage of the warrior figure whilst also illuminating the interlinking, spectral threads of gender and the sexed/gendered body in its construction.

The warrior(s)

The classical warrior

In the literature on warriors, the characters in Homer's (1991) *The Iliad* are often held up as the 'archetypal' warrior(s) and exemplars of warrior-like behaviour.[6] To paraphrase Chris Hedges, the themes of *The Iliad* include power, force, violence, and 'the everlasting fame that will be denied to [the warriors] without heroic death' (2003, p. 12).[7] Amongst the various figures within *The Iliad*, Achilles is most frequently referred to in writings about the warrior and in texts about war.[8] For Achilles, war is everything: 'it define[s] his personality and [gives] him a reason for being' and as a result *The Iliad* is primarily about Achilles' experience of war (Coker, 2013, p. 248). The argument that Achilles represents the archetypal warrior is, at least partly, predicated on his physical strength. The need for warriors of this era to exhibit a brute strength is related to the technologies (or lack thereof) of war that were available at the time. In a time when soldiers were required to slash and stab with swords or lances, to carry heavy shields, to grapple physically with one another, to use their bodily strength to climb walls, etc., physical fitness was a core component of being an able and competent soldier. Even in today's technofetishtic warfare, those roles that require the highest level of physical fitness (marines, special operations, etc.) are viewed as most closely representing the warrior ideal.

Another ancient mythical warrior is Odysseus who is described as 'an entirely new kind of warrior', he is 'quite distinct from [Achilles], [as he] uses his wits, not just his strength, to conquer his opponents' famously capturing Troy by the use of the (original) Trojan Horse (French and McCain, 2005, p. 59; Homer and Pope, n.d., p. 124).[9] Using different technology and skills from Achilles, Odysseus exhibits cunning in a stream of different escapades, including escaping from Cyclops by hiding under the body of a sheep, lying his way onto the Island of Ithaca and concealing himself as a beggar to hide from his wife (Homer and Pope, n.d.). Unlike Achilles whose key personality traits are 'loyalty, courage, honour, resistance to the misuse of political authority, independence of judgement, and raw physical power' and whose passion and physical strength render him heroic, Odysseus, whilst still mighty, prefers to use his 'guile', and his intelligence to get what he wants (Kaurin, 2014, p. 2; Coker, 2013, p. 112). What this demonstrates is that even within one iteration of the warrior, there remains difference between how that identity is constructed. Odysseus is also remembered for his return to his wife, Penelope, and for rescuing the beautiful Helen from Paris, highlighting his masculinity by protecting the women in his life, a theme that will be returned to later in the chapter under the figurations of the warrior as lover and Husband–Protector (Sjoberg and Via, 2010, p. 5). Despite their differences, Odysseus and Achilles are both courageous in the face of danger; additionally, both men are soldiers. This may seem like this should be self-evident, but being a soldier does not necessarily render the individual a 'warrior'. The former term referring to a job description and the latter to an existential and emotional connection to the experience of war, however, being a solider nevertheless provides the most obvious means through which an individual can demonstrate warrior attributes.

Brother-at-arms/blood brothers

An important relationship for the warrior was/is that with his 'brothers' in arms, for as Michael Allsep notes, 'It has largely been this shared experience of danger and deprivation in a distant place that gave combat units their invaluable comradeship' (2013, p. 387).[10] We see this not only in classical renditions of the war story, such as Achilles' connection with Patroklos, but as a leitmotif in both historical and fictional accounts of war. For example, in his St Crispin's Day Speech, Shakespeare's Henry V claims that, 'For he to-day that sheds his blood with me/Shall be my brother ...', expressing the belief that those who fight together develop a bond that is as strong as the bond between brothers (Shakespeare, 1598 Act IV, Scene III). The bond of brothers, it is worth noting, if often considered to be one of blood, i.e. their bond is based on the idea that they share the blood of their parents. This matters in relation to our thinking that 'blood is thicker than water', reflecting the idea that family relationships are the most valuable and the most deserving of our loyalty. In battle warriors may be injured and shed blood, and this shedding of blood is endowed with expressive meaning because many warriors may shed blood in pursuit or defence of the same cause.[11] Therefore, the shedding of blood in battle may be seen as the sharing of blood between the men involved, rendering them as close as brothers.

Regardless of the tools/technology employed to shed the enemies' blood, 'warriors fight and are motivated, not primarily by abstract ideals but by the man beside him' (Kaurin, 2014, p. 27). Therefore, the reason that warriors fight is not for foreign policy reasons but to save the lives of the men they serve alongside in a manner appropriate to the warrior identity. This connection with fellow men-at-arms is not a romantic fiction of 'days gone by' but something which remains pertinent today: John Bornmann acknowledged the special position of his comrades stating

Special thanks do go to the members of my team during my deployment ...
the bonds of soldiers who have served together go deeper than friendship,
and you will always be my brothers.

(2009, p. iv)

Similarly, in two accounts of individuals who served in the Royal Air Force
in the recent conflict in Afghanistan, the potential of individuals being caught in
'friendly fire' incidents was referred to as the potential for 'fratricide': the killing
of a brother (Loveless, 2010).

As noted above, the connection between brothers-at-arms is important
because it is through this that a warrior may begin to experience the expressive
side of war:

> When a warrior fights not for himself, but for his brothers, when his most
> passionately sought goal is neither glory nor his own life's preservation, but
> to spend his substance for them, his comrades, not to abandon them, not to
> prove unworthy of them, then his heart truly has achieved contempt for
> death, and with that he transcends himself and his actions touch the sublime.
>
> (Pressfield, 2000, p. 456)

As a result many autobiographical, biographical and fictional accounts of
various wars utilise the term 'Band of Brothers' to describe the close emotional
bonds between soldiers and how their lives and relationships with one another
are changed through the shared experiences of risk and loss.[12] The most well-
known of these accounts refers to the experiences of E Company, 506th Regiment, 1st Airborne in the Second World War, which was also serialised on
television. However the phrase is also used in relation to the experiences of
American soldiers in Vietnam, the US Civil War, in the US Navy in the mid-
1700s, of Australian soldiers in Vietnam and Malaysia, soldiers in the Spanish
Civil War, British Tank Regiments in the Second World War and other World
War accounts demonstrating the continuing salience of the concept in fiction and
nonfiction alike.[13] Therefore, the interaction between comrades is an important
marker for the warrior, however, the warrior is not just distinguished by his
behaviour towards his military kin but also by how he treats others.

Chivalrous gentleman

The spectral figure of the medieval knight may be found at the round table of the
legend of King Arthur, with individuals commanded to enact their roles in a very
specific way:

> Then the king stablished all his knights, and them that were of lands not rich
> he gave them lands, and charged them never to do outrageousity nor murder,
> and always to flee treason; also, by no means to be cruel, but to give mercy
> unto him that asketh mercy, upon pain of forfeiture of their worship and

lordship of King Arthur for evermore; and always to do ladies, damosels, and gentlewomen succour, upon pain of death. Also, that no man take no battles in a wrongful quarrel for no law, nor for no world's goods. Unto this were all the knights sworn of the Table Round, both old and young.

(Malory, 1485 Chapter XV)

This code illustrates that being a warrior in the form of a knight was not just about bravery. It was also predicated on particular kinds of social interactions, ideas of what constituted 'good' behaviour and these ideas, if met, constituted the basis for the knight's social position (Braudy, 2005).

The knight, as a warrior, was accorded a high status within medieval courts. Commensurate with this status he could expect to be treated in a certain way, and reciprocally was expected to behave in a certain way with particular 'facial expression, tone of voice, and manners: "magistrum referebat vultu, voce, moribus"' (Scaglione, 1991, p. 48). Braudy succinctly summarises what these expressions, manner, and tone were meant to comprise:

the chivalric values they define are loyalty (to the knight's political superior); prowess (which includes both praise of the rash willingness to throw oneself into danger ... and the skill to deal with that danger); *franchise*, or an openhanded largesse to one's fellows and followers; and courtesy to women, children, and the elderly.

(2005, p. 82)

From this list it is clear that the warrior based his status on his chivalry and as much on his etiquette in court as his prowess on the battlefield (although this remained important). This connection between military prowess and good or right social behaviours remains important in narratives of the warrior. The continuation of dining in nights at military academies and on bases, replete with dress codes, codes about how wine or port should be passed, poured and drunk, and codes about how men and women are expected to interact indicate that counter to suggestion, chivalry is not dead.

The chivalry of an individual in the Middle Ages (the heyday of this kind of warrior) not only depended on how he behaved, but also on *who* he was, revealing the ghostly shape of class privilege (Braudy, 2005, p. 65). The identification of the warrior as chivalrous was more likely to be bestowed on individuals from higher classes (Scaglione, 1991). For example, Braudy notes, 'As the perception of masculinity developed in relation to war in the Middle Ages, "knight" was the bridge between the otherwise separable or even contradictory categories of "noble" and "warrior"' (2005, p. 71). This connection may have resulted from the cost of enacting the warrior role. Costs such as maintaining stables and the associated staff, armour, and training, as well as access to the kind of educational environment that provided insight into the necessary etiquette were almost always only available to the very wealthy, and in the historical European context this meant the landed gentry.[14]

The chivalrous knight (coded masculine) was frequently constructed in opposition to the barbarian (coded feminine and dangerous) 'Other', and therefore is often racialised as white.[15] Graham Dawson traces the ghosts of chivalry within English national identity through colonial endeavours, noting that in the 1800s 'India was the perfect setting for chivalric adventure', one where fortunes could be gained through conquest, exploration into unknown 'native' areas could be undertaken, and a man could prove his masculinity through the subduing the local populace (Dawson, 1994, p. 60). In the narratives of the time, subduing of the local populace might be understood as necessary for the 'betterment' of that populace, overlaying the *colonial* knight with patriarchal discourses. Violence against the natives might then be articulated as being 'for their own good', feminising and infantilising the local population in the face of the fatherly figure of the white knight. Reflecting this racialised and colonial background, a chivalrous knight might attempt to bring religion (usually in the form of Christianity) to the 'heathen' natives (Braudy, 2005, pp. 289–292). This character would not only seek to evangelise but also be modest about his successes, a history of which equates the term 'Christian' not only to religious belief but also to a clean living and well-mannered individual. The warrior then is implicated in the pursuit of what is 'right' and what is 'good'. His military actions should be framed in terms of this language and this is part of how he makes sense of the existential element of warrior-hood.

Discussions of evangelising knights would be incomplete without reference to the knights involved in the crusades. The Order of the Poor Knights of the Temple of Solomon, more commonly known as The Templars, was initially founded to protect pilgrims in Jerusalem from robbery and muggings, but they were suddenly catapulted to fame and power in 1129 by a series of papal bulls.[16] Dedicated to the doomed cause of the Christian crusades against the Muslim Saracens in the Middle East, the Templars '[took] vows of chastity, observe[d] discipline at home and on the battlefield, [ate] in silence, and [held] everything in common' (Maxwell-Stuart, 2006, p. 96). The aim of these vows was to establish The Templars as different from previous groups of knights by rejecting worldly goods and practising a monastic way of life whilst remaining militaristic and armed.[17] The connection between a religious living and militarism can also be seen in the Buddhist monks of the Shaolin Temple in China. As part of their training, these monks undertook 'a series of highly demanding physical exercises … designed to help the monks focus and control their chi' the aim of which was to enable them to defend their temple and also to attain enlightenment (French and McCain, 2005, p. 188). These two aims neatly connect the idea of the physical/instrumental and the existential/expressive elements associated with the figure of the warrior.[18] That is, the brute strength required to be successful in hand-to-hand combat, and the emotional and psychological courage to be prepared to die for the cause or 'find oneself' through the experience of war. The Shaolin warrior monks were also strictly guided as to the use of violence which they could not use for any material gains, nor for 'ego-driven goals like fame or glory' (French and McCain, 2005, p. 192). Unlike the Knights Templar, the

Shaolin monks did not fight to evangelise or eradicate non-believers, but only in self-defence – or the defence of others (French and McCain, 2005, p. 193).

The Samurai of Japan also represented an incarnation of the chivalrous gentleman. So closely did the European medieval knight and Samurai resemble one another that, Ehrenreich claims, 'had [they] met on the same road, they would have immediately recognized each other as kin' (1998, p. 144). Like the European medieval knight, the Samurai's actions and behaviour were governed by a series of rules and codes 'touching every aspect of the warrior's existence, from the sacred to the mundane' (French and McCain, 2005, p. 199). The expressive component of warfare was of particular importance to the Samurai: 'death had meaning' because it enabled the individual to demonstrate 'honour, courage and loyalty' all things that enabled the Samurai to also make sense of life (Coker, 2002, p. 7). Informed by Shinto, Buddhism, Taoism, and Confucianism, the Bushido code ('The Way of the Warrior') was imbued with layers of transcendental meaning connecting the use of violence with spiritual significance and the virtue of honour.[19] The ghost of the Samurai was so potent that Kamikaze pilots in the Second World War were deemed to be channeling this figure hundreds of years later and through the modern medium of the fighter plane.[20]

The beautiful hero

In addition to his brothers-at-arms, the warrior had a special relationship with another, one which was connected to chivalrous behaviour. In many accounts the warrior is not framed as demanding the love of a woman, but as warranting it, not only through his heroic deeds, but also through his physical attractiveness.[21] In Pre-Raphaelite renditions of the warrior, particularly as a knight, the figure is one of exquisite masculine physique, aristocratic features, and fine clothes (see, for example, Edmund Blair Leighton, Dante Gabriel Rossetti, John William Waterhouse, and earlier examples by Carpaccio Vittore and Pieter Bruegel). No matter what the context, the warrior, in art, is never ugly. For example, in painted Ancient Greek ceramics, the warrior's body is muscled and strong, whether wrestling fearsome monsters with his bare hands or engaged in discussions with fellow warriors, his biceps and thigh muscles are bulging and shapely, his pectorals carefully inscribed on his skin. His beard is neatly trimmed, and his body adorned with swirls, and symbols describing a body decorated with jewellery. Additionally, as a reminder of the warrior's appeal to moral 'goodness' many paintings of warriors render them saint-like, their heads adorned with brightly coloured or golden halos, highlighting their status as (almost) superhuman/godlike. Additionally, it is not only the warriors who are depicted as beautiful but the weapons the warriors display are nearly as beautifully rendered as the warriors themselves: jewel encrusted swords, highly decorated shields, and complex, detailed (and undoubtedly extremely heavy) armour, some endowed with magic or superpowers preventing their wearer's/wielder's death or injury.[22] As the sword of Excalibur provides an example, sometimes the weapons are as or more well known than the individuals who utilise them in their chivalrous deeds and duels.[23]

The individuals who wield these magical weapons are (usually) warriors who are involved in hand-to-hand combat, where physical strength and skill is paramount.[24] In this close killing environment the warrior's bravery and existential 'goodness' is tested as he must look his opponent in the eye as he tests his body against his enemy (Coker, 2007, p. 18). Instrumentally conceived, '[t]he hardened body of the soldier warrior is now a real or potential weapon ...', it may be 'envisaged as a permanently connected, interoperable and flexible "platform" for the delivery of force ...' (McSorley, 2013, p. 10; Whitworth, 2007, p. 161). The soldier's body is the material through which war is waged, 'a crucial foundation', it is the physical presence on the war front and the means through which the force which comprises war is enacted (Lande, 2007, p. 96). Although the physical body of the soldier risks damage, dismemberment and death during war time, the ghostly figure of the warrior, as a masculine (rather than male) subject is constructed as *in*vulnerable. Using Gaten's understanding of the body as not just as the physical (although this is important too) but also symbolic (imaginary) we can see how the vulnerability of the soldier and the invulnerability of the masculine can co-exist, illustrating one place in which the queer logic outlined in the previous chapter is expressed.[25]

The symbolic body of the soldier is important because it is through the identity of the warrior that the soldier understands his role in war: what constitutes 'good' reasons to utilise violence, and how the ghostly warrior of multiple pasts can be translated into the modern situation and understood as a masculine role model. The physical body of the solider is important because it is through this physical body that war is fought, it is both the means through which violence is inflicted, and the place where the results of that violence might be transcribed in scars and wounds. Therefore, whilst the figure of the warrior is important as a symbolic body, how this symbolic body is understood has implications for the real, live, fleshy bodies of soldiers.

Warrior as lover

In many myths, legends and stories the warrior's relationship with his (usually, but not exclusively female) lover is one of his defining characteristics and reasons for bearing arms. The need to defend and love women is a repeating theme through much of the literature on warriors, particularly echoing through the idea of 'courtly love'. This concept may be defined as 'staging vexed heterosexual liaisons between aristocratic couples within an idealized public sphere of refined court life in the High Middle Ages ...' (Burns, 2001, p. 23). Courtly love was expressed through a knight's devotional actions towards his love interest in the form of carrying 'favours' from her (i.e. a handkerchief or flower) into battle, challenging anyone who offended her, and writing poetry or songs about her.[26] Edward Leighton's painting *God Speed* depicts an example of courtly love as a woman ties a red sash around the arm of the mounted knight as a symbol of her regard for him and as an object for him to return to her when he comes back, safe and well, from heroic battle. Courtly love was generally held to be unrequited and it was considered perfectly reasonable for the object of the courtly love to be married to another man (see, for example, the rules laid out by Cappellanus in 1184). Whilst it was considered suitable for a warrior knight to love a woman from afar, to indulge in 'the excess to which their passions might naturally tempt them' was something to be fought against (Lash, 1995, p. 25). The temptations of the flesh in its fully carnal form was considered unworthy of the chivalrous knight, who should love his lady with 'loving esteem, worthiness and ... dignity' (Lash, 1995, p. 25). As Braudy notes 'the lure of love, expressed through the twin enticements of the sexual and the domestic, has the tendency to erode the warrior ethic and expose the knight to potentially subversive forces', acting as a reminder of the dangers associated with the feminine in relation to the masculine warrior identity (2005, p. 106).

The warrior, however, is not entirely immune from the connection between sexual prowess and physical masculinity. The importance of the body to the Warrior as Lover can be seen in the following quote:

> His hold on life was marvellous. He didn't die, and the bits seemed to grow together again. For two years he remained in the doctor's hands. Then he was pronounced a cure, and could return to life again, with the lower half of his body, from the hips down, paralysed for ever ... The gay excitement had gone out of the war ... dead.
>
> (Lawrence, 2013, pp. 7–9, 117–119)

Taken from *Lady Chatterley's Lover*, without a fully functioning body the individual is unable to enjoy the 'gay excitement' of war or that of conjugal relations with his wife. The extensive martial trauma causes extensive *marital* trauma, and his wife takes a lover. This connects the physical incapability of the wounded soldier to perform his role as a warrior and his role as a husband: both roles relying heavily on the physical and sexual performance of masculinity. In their analysis of why soldiers rape in war, Maria Eriksson Baaz and Maria Stern (2009) note that soldiers in the Congo (DRC) identified rape because the men had sexual needs that were not being fulfilled as not 'evil' rape, but rather as something unfortunate but understandable. Importantly, they note that 'Beneath these accounts underlies an understanding of men's (hetero)sexuality as a driving force, which, when unleashed by the climate of warring in which "normal" societal controls are suspended and the rules of warfare reign, easily results in rape' at least in part because warrior masculinity is associated with specific and aggressive performances of masculinity (Baaz and Stern, 2009, p. 497).

Husband–Protector

> For thousands, probably tens of thousands of years, we men have laid down our lives so that the women we love might live.
>
> (Van Creveld, 2013)

As shown in the story of St George and the Dragon, on completing the onerous and dangerous task set before him, many a warrior either returns to or is awarded with a woman, usually a wife.[27] In addition to being a 'gift' bestowed on the warrior in recognition of his manly heroism, the 'wife' is an individual whom the warrior is bound to protect, and this becomes a marker of his identity.[28] Within the traditional family unit the husband/male is usually described as the head of the household, with the wife/female and children as subservient to him.[29] Where his home (read: wife and children) are endangered, the warrior is duty-bound to provide protection through his physical prowess and martial skill.[30] Arguments about the masculine 'urge' to protect, Charlotte Hooper argues, are ultimately connected with ideas of the 'naturalness of men's aggression' so that protection and masculinity are constructed as mutually constitutive (Hooper, 2001, p. 81). As the Van Creveld (2013) quote above illustrates, the importance of having someone 'at home' as a motivator to protect continues to haunt our opinions of war and the appropriate activities for individuals in war.[31]

The Husband–Protector's 'urge' to protect his family represents a specific iteration of masculinity, one that appears in some ways to contradict that of the warrior. As Iris Marion Young notes, the protector is a

> benign image of masculinity, one more associated with ideas of chivalry [where] real men are neither selfish nor do they seek to enslave or over-power others for the sake of enhancing themselves. Instead, the gallantly masculine man is loving and self-sacrificing, especially in relation to women. He faces the world's difficulties and dangers in order to shield women from harm and allow them to pursue elevating and decorative arts. The role of this courageous, responsible, and virtuous man is that of a protector.
>
> (2003, p. 4)

This iteration of the warrior is not focused on killing the enemy, but on risking his life to save the lives of those his role and identity as a warrior require him to protect.[32] Laura Sjoberg uses the dedication to Paul Ray Smith in the *St Petersburg Times* as an example of how the ideal warrior is constructed as a 'husband, soldier, and protector':

> We can remember a husband who cherished his wife. We can praise a father who loved his children. We can recall a soldier who cared for his men. And we can celebrate the life and legacy of a man who gave the last full measure of devotion to his country.
>
> (Cited in Sjoberg, 2010a, p. 214)

Paul Ray Smith's death is then constructed through this and other dedications as the death of an honourable warrior. He has died to save the lives of those he loves, to protect his family (wife and children) and his nation ('the motherland'). Jean Bethke Elshtain (1995) theorised this reading of the masculine warrior as 'the Just Warrior' who aims to protect the 'Beautiful Soul' who is weak, virginal and incapable of protecting herself (and therefore without agency).[33] We can see this logic haunting modern day conflicts, as Lee notes, 'From Obama to Blair, establishing and maintaining support for military interventions has been pursued in part by repeatedly stressing the need to protect women ...' (2015, p. 99). Such an approach brings to the forefront not just the gendered logic of the male protector, but also the racialised logic of the white man protecting brown women from brown men.[34] This idea of white masculinity as 'saviour' of brown women has been illuminated in relation to the recent US interventions in Iraq and Afghanistan. Laura Shepherd (2006) showed how the previously invisible Afghan woman suddenly began to show up in the media, always as a victim, always denied of agency and always in need of emancipation from the decent, white, masculine, American soldier. As such, the white male warrior in the form of the 'English Gentleman' is positioned as the protector of both white and brown women from the oriental savagery of brown men who place women at risk of being harmed.[35]

The concept of the manly protector hinges on the idea that the feminine is in need of that protection (and specifically the protection of *white* masculinity). Critics of women's participation in the military are haunted by the fear that men might 'get themselves killed' trying to protect their female comrades (although this ignores the protective impulse of the brother-at-arms narrative).[36] Because the woman has been traditionally coded as needing protection, these critics argue, the male soldiers will 'naturally' feel more inclined to come to the aid of the female comrades even when it is dangerous and against their orders to do so. Additionally, the warrior's identity *requires* that there be someone 'at home' to protect, as without 'the protected', the warrior loses both part of his identity and his source of reward.[37] Allowing women onto the battlefield deprives the warrior-protector of his rationale for risking his life, and some argue, degrades the entire experience of war.[38] As a result the theatre of war is separated from the 'home front', which clearly delineates the differentiation between the public and private, and therefore the spheres that belong to the masculine and the feminine.[39] For example:

> The role of the military spouse in the modern age is indicative of this dichotomization. Military spouses (read: wives) are socialized to accept their role as caretaker and peacemaker, and further socialized to accept that these issues are secondary issues, a footnote to the 'national security interest'.
>
> (Horn, 2010, p. 62)

As well as limiting the agency of women in war time, feminising the country or nation, or personifying it as a woman, meant that warriors could play out the role of protector on a larger scale.[40] The nation is constructed simply as a larger version of the 'standard household' whereby the husband is required to protect against external threats and to direct the behaviour of the women and children (his subordinates). As Iris Marion Young notes:

To the extent that citizens of a democratic state allow their leaders to adopt a stance of protectors toward them, these citizens come to occupy a subordinate status like that of women in the patriarchal household.... At the same time that it legitimates authoritarian power over citizens internally, the logic of masculinist protection justifies aggressive war outside.

(Young, 2003, p. 2)

In addition to gendering and racialising the role of the warrior protector, the casting of the warrior as 'protector' establishes his heteronormativity. There is no place in this formulation for the warrior who is protecting his (male) partner at home. There is also no place for the female warrior to be protecting her husband or (female) partner. The armour provided for the husband–protector is one-size fits all: that is, designed to fit the hard, erect male physique with a space for a keepsake, a remainder of the woman left behind, in need of his protection.

Reflections on masculinity and the figure(s) of the warrior

Whilst the literature on the warrior, and the social and cultural artefacts that reflect its power, is/are wide ranging and varied, there is little acknowledgement of the overarching theme of masculinity that pervades all of the commentary (with the exception of Braudy, 2005). Even in texts that include references to female warriors these individuals are constructed as 'the Other' and as anomalies that prove the rule.[41] Masculinity, in the discourses of the warrior, acts as an implicit organising principle in that it shapes what can and cannot be construed as warrior-like behaviour (Peterson, 2010). For example, the masculinity of the Classical Warrior is largely constructed around his physical prowess. In the worlds inhabited by warriors like Achilles and Odysseus, battle relied on the strength of each individual to lift and adequately use the basic equipment they were given (primarily swords or spears and shields). Using these weapons, the soldiers were required to demonstrate their identity as warriors by the piercing of the enemies' flesh, or using their bare hands to kill. Those who could kill in this manner, and kill many, were granted the highest status and deemed the best representation of manliness/masculinity, and therefore the best warriors.[42]

To demonstrate strength was to demonstrate masculinity, and correspondingly to be weak was to be feminine. A warrior's strength was understood only in opposition to weakness. Therefore, that meant that the warrior needed not just to be strong but *stronger than* others, demonstrating the relationality of not just strength and weakness but their relationship to masculinity and femininity. In addition to 'raw physical power', the masculinity of the Classical Warrior is established through action and deployment of intellect and judgement (Kaurin, 2014, p. 2). As noted in Chapter 1, action is situated in opposition to passivity in a binary. Intellect and judgement are informed by the binaries rational/emotional and mind/body. In both instances, as with active/passive, the first term is associated with the masculine and is constructed as superior to the second term which is associated with the feminine. By being cunning and using his guile to get out of difficult situations, Odysseus enacts the rationality associated with masculinity.

The 'Brother-at-Arms' iteration of the warrior illuminates the power of friendships in times of trial that rendered the homosocial bond uniquely strong (and inaccessible to women) and enabled the maintenance of the myth of hetero-sexuality.[43] The 'brothers' within the military provide the audience to whom the warrior may (indeed must) perform his masculinity. This may be through the performance of the masculine component of gendered binaries such as strength (as opposed to the feminine weakness) or through other attributes associated with the masculine (rationality through the suppression of emotion, or focus on the cultural rather than the 'natural').[44] The masculinity of the warrior in the 'Brother-at-Arms' iteration may also be partly attributed to the larger numbers of men than women in the vast majority of armed forces. In some instances, contemporary resistance to the inclusion of women is framed around the fear of disrupting the bond between the brothers-at-arms through the introduction of 'sisters'.[45]

The 'Chivalrous Gentleman' constructs his identity by providing support to the frail flowers of femininity who lacked the agency of the masculine knight.[46] Similarly, the 'white/European' warrior masculinity is situated as hierarchically superior to the racialised barbarian 'Other' who is infantilised as needing to be 'civilised'. The process of 'civilising' reinforces the masculinity/femininity hierarchy, situating the civiliser as the masculine, active role against the feminised (and infantilised) 'Other'. The Chivalrous Gentleman renders explicit the connection between the existential component of war and patriarchal constructions of masculinity through its relationship with crusading and evangelic European Christianity.[47]

The 'Beautiful Hero' connects the idea of moral 'goodness' with physical attractiveness, so that masculinity is constructed as being physically (and sexually) appealing to the opposite sex. The physical strength of the warrior is, in this iteration, not just about winning wars (although this remains integral to his identity) but rather strength, and the embodiment of it, becomes oppositional to the feminine by reinforcing heterosexualism. Similarly, the capacity of the beautiful hero to 'thrust' his sword blade deep into the enemy reinforces the gendered and sexualised binary of the penetrating and the penetrated.

A similar binary logic is displayed through the Warrior as Lover where the warrior is sketched out as unfulfilled lover (avoiding dangerous encounters with the feminine lover) in preference for the cerebral (rational?) love. However, this unrequited love as masculine only functions when the warrior has the physical/ sexual capacity to fulfil his desires but chooses not to. The broken body of the Lady Chatterley's husband and her cuckholding of him by taking a lover serves to demonstrate the importance of the physical (and sexual capability) of the warrior's body.[48] The emasculation of Lady Chatterley's husband is compounded by the Husband–Protector iteration of the warrior. In this iteration the masculinity of the warrior is clearly created by his capacity (and responsibility) to protect the home/domestic from the ravages of war, carefully delineating between the two spheres and situating the feminine in the former, rendering the war sphere particularly and peculiarly masculine. The 'Just Warrior' must protect the 'Beautiful Soul', and as a result the 'Beautiful Soul' must submit to being protected and to being banished from the theatre of war, reinforcing both gendered discourses and patriarchal heteronormativity.[49]

Conclusion

This chapter considered the importance of the 'warrior' to discourses of war. The warrior is an inherently masculine figure, who I have sketched in this chapter as unproblematic, providing the backdrop to the critical engagements with the concept in the following chapters. With reference to Chapter 2, I argue that the warrior is a ghostly figure, one that is unstable, real whilst not real. As such, the chapter has traced some of the different iterations of the warrior, illustrating the cross-hatching of the spectral and the way that this creates something that is coherent whilst also being ghostly. There are many ways in which history, myth, fiction, fact, art, legend, and popular culture have differently engaged with the idea of the warrior, and the way in which the ghostly warrior has a wide and varied parentage in no way inhibits its coherence.

The 'traditional' narratives of the warrior that I have described in this chapter do not represent the entire collection of warrior stories, but rather a select reading that has tried to indicate some of the many differences between and within warrior-types, across a range of different historical, cultural, and geographical contexts. The ghost hunt reveals some of the wide range of histories, myths, pieces of art, poetry, and pieces of fiction which have informed the development of the figure of the warrior. Whilst it is possible to call to mind 'a' warrior, this chapter has demonstrated that 'he' comes in many different guises, that there is no one quintessential warrior but many possible warrior figures. As noted in the chapter introduction, this collection is not intended to provide a representation of all of the possible iterations of the warrior but provide an interesting sample for exploration. In this ghost hunt, the ghosts of the Classical Warrior, the Brother-at-Arms, the Chivalrous Gentleman, the Beautiful Hero, the Warrior-as-Lover, and the Protector–Husband have layered over one another to create the simultaneously singular and yet always changing figure of the warrior.

Within the variations, this chapter has traced the implicit masculinity which haunts each of the iterations. The warrior, regardless of how s/he is conceived, is always enacting the masculine role that is constructed in opposition to the feminine.[50] Whether the feminine is represented by a relationship with physical women or by inferior traits in gendered binaries, the masculinity of the warrior relies on his oppositional relationship to the feminine.[51] In keeping with the commitment of Haunting to render explicit forms of abuse and power, this chapter has outlined how the masculinity of the warrior places him in a position of power over the feminine. This hierarchy enables the warrior to act in ways that would be constructed as inappropriate for feminine characters, or in some instances enables the warrior to act where the feminine character is not permitted to act at all; demonstrating the power of the active/passive binary in gendered discourses.[52] Masculinity also constrains the actions of the warrior – he must act in ways commensurate with his masculine status or risk losing it.[53]

As noted previously, this chapter introduces the methodology of the ghost hunt described in the previous chapter and painted the backdrop of the warrior figure who is implicated in the lives and experiences of the British Reaper crews who form the focus of the subsequent chapters. The warrior figure has survived, through mutation and the romance associated with the ghost, the introduction of various different military technologies – from the gun, to the aeroplane, to the A-bomb, and remains a potent influence on the members of the British military today. The interaction with these technologies has required the privileging of different kinds of masculinity, be that through 'ferocious aggressiveness' required for hand-to-hand combat or the 'passive disdain', the calm, calculating rationality needed by the sniper (Ehrenreich, 1998, p. 9). Following on from this, the subsequent chapters move to consider how the warrior and his attendant masculinity is challenged and reinforced by the introduction of drone technology.

Notes

1　(Van Creveld, 2001; Coker, 2007a, 2013).
2　(Coker, 2013; Kaurin, 2014; French and McCain, 2005).
3　(See for example Van Creveld, 2001; Coker, 2013; Elshtain, 1995; Sylvester, 2013; Pressfield, 2011; French, 2005).
4　(Ehrenreich, 1998; Coker, 2002; French and McCain, 2005).
5　As a second caveat to this chapter, I have chosen these iterations because they have particular cultural resonance for the case study of the thesis. Therefore, whilst the chapter is not *exclusively* concerned with Western warrior ghosts, it is certainly Western-centric in its focus. As such, I reiterate, this collection of ghostly warriors is not intended to be comprehensive, but rather to serve as a useful backdrop to the discussions in the subsequent empirical chapters, which focus on the experiences of British servicemen and women.
6　(Van Creveld, 2001; Coker, 2001; Pressfield, 2011; French, 2005). Although one of the aims of this chapter is to destabilise this claim that Achilles is *particularly*, or more archetypal than any of the other iterations described here.
7　And how this relates to apparently riskless drone crews is discussed in detail in Chapter 4.
8　(Freedman, 2013; Scheipers and Greiner, 2014; Hartsock, 1984; Damousi and Lake, 1995; Norris, 2008).
9　Odysseus is also referred to as Ulysses.
10　An important theme for Reaper crews that is unpicked in Chapter 6.
11　(Kaurin, 2014)
12　(For example, Fleming, 1996; McDonald, 1989; Brune, 2000; Winters and Kingseed, 2011; Urban, 2014).
13　(Fleming, 1996; McDonald, 1989; Winters and Kingseed, 2011; McFarlane and James, 2010; Brune, 2000; Urban, 2014; Arthur, 2009).
14　(Lambdin, 1999; Chibnall, 1991).
15　(Maxwell-Stuart, 2006; Tickner, 1992; Hooper, 2001).
16　(Martin, 2011).

17 (Napier, 2011a, 2011b).
18 (Kaempf, 2014, 2009; Coker, 2013).
19 (French and McCain, 2005, p. 200).
20 (Bays, 2008; Coker, 2002).
21 (Francis, 2008; Homer and Pope, n.d.; Scaglione, 1991).
22 (Malory, 1485; Knowles and Speed, n.d.).
23 (Malory, 1485; Knowles and Speed, n.d.).
24 (Allsep, 2013; Stephenson, 2012; Kaurin, 2014; Ehrenreich, 1998).
25 (Gatens, 1996; see also Golden, 2006).
26 (Burns, 2001).
27 (Gale, 1777; Lash, 1995).
28 (Hicks Stiehm, 1982; Carver, 2006).
29 (Peterson and Runyan, 2010).
30 (Pin-Fat and Stern, 2005; Elshtain, 1995; Cooke and Woollacott, 1993; Stern and Zalewski, 2009).
31 (Elshtain, 1995; Van Creveld, 2001; Cohn, 2000).
32 (Elshtain, 1995).
33 (Elshtain, 1995; See also Sjoberg, 2007, p. 4; Cooke, 1996).
34 (Spivak, 1988, p. 297; Duncanson, 2014; Sjoberg and Via, 2010).
35 (Hooper, 2001; Dawson, 1994).
36 (Van Creveld, 2000; Morgan, 1994).
37 (Pin-Fat and Stern, 2005).
38 (Van Creveld, 2000, 2001).
39 (Horn, 2010; Via, 2010).
40 (Goldstein, 2001; Braudy, 2005, p. 22).
41 (Goldstein, 2001; Van Creveld, 2001; Mayor, 2014).
42 (Homer and Fagles, 1991; Coker, 2014).
43 (Welland, 2013).
44 (Hooper, 1998; Duncanson, 2007).
45 (Klimas, 2015; Mitchell, 1989; Caforio, 2006; Carreiras, 2006).
46 (Malory, 1485; Geoffrey Chaucer, n.d.; Burns, 2001).
47 (Dawson, 1994; Napier, 2011a).
48 (Lawrence, 2013).
49 (Elshtain, 1995; Peterson, 2014b, 2014a).
50 (Elshtain, 1995; Braudy, 2005).
51 (Grosz, 1994; Peterson and True, 1998).
52 (Elshtain, 1995; Pettman, 1996).
53 (Carver, 2006).

References

Allsep, L.M. (2013) The myth of the warrior: martial masculinity and the end of don't ask, don't tell. *Journal of Homosexuality*, 60 (2–3): 381–400. doi: 10.1080/00918369.2013.744928.

Arthur, M. (2009) *The real band of brothers: first-hand accounts from the last British survivors of the Spanish Civil War*. London: Collins.

Baaz, M.E. and Stern, M. (2009) Why do soldiers rape? Masculinity, violence, and sexuality in the armed forces in the Congo (DRC). *International Studies Quarterly*, 53 (2): 495–518. doi: 10.1111/j.1468-2478.2009.00543.x.

Basham, V. (2013) *War, identity and the liberal state: everyday experiences of the geopolitical in the Armed Forces*. Interventions. 1st edn. London; New York: Routledge, Taylor & Francis Group.

Bays, C. (2008) *Norms amidst chaos: chivalry and rational choice in combat.*

Bornmann, J.W. (2009) *Becoming soldiers: army basic training and the negotiation of identity.* The George Washington University.

Braudy, L. (2005) *From chivalry to terrorism: war and the changing nature of masculinity.* 1. Vintage Books ed. New York: Vintage Books.

Brune, P. (2000) *We band of brothers: a biography of Ralph Honner, soldier and statesman.* St Leonards, NSW: Allen & Unwin.

Burns, E.J. (2001) Courtly love: Who needs it? Recent feminist work in the medieval French tradition. *Signs,* 27 (1): 23–57.

Caforio, G. (2006) *Handbook of the sociology of the military* [online]. New York, NY: Kluwer Academic/Plenum Publishers. Available at: http://site.ebrary.com/id/10189064 (accessed: 13 October 2015).

Carreiras, H. (2006) *Gender and the military: women in the armed forces of Western democracies.* London; New York: Routledge.

Carver, T. (2006) Being a man. *Government and Opposition,* 41 (3): 450–468.

Chaucer, Geoffrey (n.d.) *The Canterbury tales.* Laing Purves, D. (ed.). Tustin, CA: Xist Classics.

Chibnall, M. (ed.) (1991) *Anglo-Norman studies XIII: proceedings of the Battle Conference 1990.* Woodbridge: Boydell.

Cohn, C. (2000) 'How can she claim equal rights when she doesn't have to do as many push-ups as I do?': the framing of men's opposition to women's equality in the military. *Men and Masculinities,* 3 (2): 131–151.

Coker, C. (2001) *Humane warfare.* London; New York: Routledge.

Coker, C. (2002) *Waging war without warriors?: the changing culture of military conflict.* IISS studies in international security. Boulder, CO: Lynne Rienner Publishers.

Coker, C. (2007) *The warrior ethos: military culture and the war on terror.* LSE international studies series. London; New York: Routledge.

Coker, C. (2013) *Warrior geeks.* London: C. Hurst & Co. Publishers Ltd.

Coker, C. (2014) *Men at war: what fiction tells us about conflict, from the Iliad to Catch-22.* London: C. Hurst & Co. Publishers Ltd.

Cooke, M. (1996) *Women and the war story.* Berkeley, CA: University of California Press.

Cooke, M. and Woollacott, A. (1993) Gendering war talk [online]. Available at: http://alltitles.ebrary.com/Doc?id=10899124 (accessed: 29 July 2015).

Damousi, J. and Lake, M. (eds) (1995) *Gender and war: Australians at war in the twentieth century. Studies in Australian history.* Cambridge: Cambridge University Press.

Dawson, G. (1994) *Soldier heroes: British adventure, empire, and the imagining of masculinities.* London; New York: Routledge.

Duncanson, C. (2007) *Forces for good? British military masculinities on peace support operations.* Edinburgh: AIAA.

Duncanson, C. (2014) Masculine rivalries and security: the US and UK in Iraq and Afghanistan. *E-International Relations* [online], 4 July. Available at: www.e-ir.info/2014/07/04/masculine-rivalries-and-security-the-us-and-uk-in-iraq-and-afghanistan/ (accessed: 19 May 2015).

Ehrenreich, B. (1998) *Blood Rites.* St Ives, Great Britain: Virago.

Elshtain, J.B. (1995) *Women and war.* University of Chicago Press ed. Chicago, IL: University of Chicago Press.

Francis, M. (2008) *The flyer: British culture and the Royal Air Force, 1939–1945.* Oxford; New York: Oxford University Press.

Freedman, L. (2013) *Strategy: a history*. Oxford; New York: Oxford University Press.

French, S. (2005) Warrior transitions: from combat and social contract. JSCOPE [online]. Available at: http://isme.tamu.edu/JSCOPE05/French05.html

French, S.E. and McCain, J. (2005) *The code of the warrior: exploring warrior values past and present*. 1. paperback edn. Lanham, Md.: Rowman & Littlefield.

Gale (1777), St George and the dragon. Eighteenth Century Collections Online. Available at: http://find.galegroup.com/ecco/retrieve.do?scale=0.33&docLevel=FASCIMILE&prodId=ECCO&tabID=T001&resultListType=RESULT_LIST&retrieveFormat=MULTIPAGE_DOCUMENT&inPS=true&userGroupName=bham_uk&docId=CB3326873555&currentPosition=1&workId=1701401100&relevancePageBatch=CB126873555&contentSet=ECCOArticles&callistoContentSet=ECCOArticles&resultListType=RESULT_LIST&reformatPage=N&retrieveFormat=MULTIPAGE_DOCUMENT&scale=0.33&pageIndex=2&orientation=&showLOI=&quickSearchTerm=&stwFuzzy=&doDirectDocNumSearch=false

Gardiner, Steven (2012) The warrior ethos: discourse and gender in the United States Army since 9/11. *Journal of War and Cultural Studies*, 5 (3): 371–383.

Goldstein, J.S. (2001) *War and gender: how gender shapes the war system and vice versa*. Cambridge: Cambridge University Press.

Golden, K.B. (2006) *Nietzsche and embodiment: discerning bodies and non-dualism*. SUNY series in contemporary continental philosophy. Albany: State University of New York Press.

Gatens, M. (1996) *Imaginary bodies: ethics, power, and corporeality*. London; New York: Routledge.

Grosz, E.A. (1994) *Volatile bodies: toward a corporeal feminism. Theories of representation and difference*. Bloomington: Indiana University Press.

Hartsock, N.C.M. (1984) Masculinity, citizenship, and the making of war. *PS: Political Science & Politics*, 17 (02): 198–202.

Hedges, C. (2003) *War is a force that gives us meaning*. 1. Anchor Books edn. New York: Anchor Books.

Hicks Stiehm, J. (1982) The protected, the protector, the defender. *Women's Studies International Forum*, 5 (3/4): 367–376.

Homer and Fagles, R. (1991) *The Iliad*. Harmondsworth: Penguin.

Homer and Pope, A. (n.d.) *The Odyssey*. Kindle version.

Hooper, C. (1998) Multiple masculinities in international relations. In Zalewski, M. (ed.) *The 'man question' in international relations*. Boulder, CO: Westview Press.

Hooper, C. (2001) *Manly states: masculinities, international relations, and gender politics*. New York: Columbia University Press.

Horn, D.M. (2010) Boots and bedsheets: constructing the military support system in a time of war. In Sjoberg, L., Via, S., and Enloe, C. (eds) *Gender, war and militarism: feminist perspectives*. Santa Barbara, CA: Praeger. pp. 57–68.

Kaempf, S. (2009) Double standards in US warfare: exploring the historical legacy of civilian protection and the complex nature of the moral-legal nexus. *Review of International Studies*, 35 (03): 651.

Kaempf, S. (2014) Postheroic U.S. warfare and the moral justification for killing in war. In Gentry, Caron E. and Eckert, Amy E. (eds) *The future of just war: new critical essays*. Athens: Georgia University Press. pp. 79–97.

Kaurin, P.M. (2014) *The warrior, military ethics and contemporary warfare: Achilles goes asymmetrical*. Military and defence ethics. Farnham, Surrey: Ashgate Publishing Company.

Klimas, J. (2015) Integrating women into combat reduces effectiveness, harms unit cohesion. *Washington Times* [online], 19 March. Available at: www.washingtontimes.com/news/2015/mar/19/problems-women-combat-cant-be-mitigated-report/?page=all (accessed: 13 October 2015).

Knowles, J. and Speed, L. (n.d.) *The legends of King Arthur and his knights*. Hamburg, Germany: Tredition.

Lambdin, L.C. (ed.) (1999) *Chaucer's pilgrims: an historical guide to the pilgrims in The Canterbury tales*. Westport, CT: Greenwood Press [u.a.].

Lande, B. (2007) Breathing like a soldier: culture incarnate: breathing like a soldier: culture incarnate. *The Sociological Review*, 55: 95–108. doi: 10.1111/j.1467-954X.2007.00695.x.

Lash, J. (1995) *The hero: manhood and power*. Art and imagination series. New York: Thames and Hudson.

Lawrence, D.H. (2013) *Lady Chatterley's lover*. A public domain book.

Lee, P. (2015) *Truth wars: the politics of climate change, military intervention and financial crisis*. London: Palgrave Macmillan.

Loveless, A. (2010) *Blue sky warriors: the RAF in Afghanistan in their own words*. Sparkford, Yeovil, Somerset, UK: Haynes Publishing.

Malory, T. (1485) *Le morte d'Arthur*. Project Gutenberg. Available at: www.gutenberg.org/files/1251/1251-h/1251-h.htm (accessed: 14 July 2015).

Martin, S. (2011) *The Knights Templar* [online]. New York: Oldcastle Books. Available at: http://public.eblib.com/choice/publicfullrecord.aspx?p=845146 (accessed: 24 July 2015).

Maxwell-Stuart, P.G. (2006) *Chronicle of the popes: the reign-by-reign record of the papacy from St Peter to the present*. 2nd edn. London; New York: Thames & Hudson.

Mayor, A. (2014) *The Amazons: lives and legends of warrior women across the ancient world*. Princeton, NJ: Princeton University Press.

McDonald, W. (1989) *A band of brothers: stories from Vietnam*. Lubbock, TX: Texas Tech University Press.

McFarlane, B. and James, D. (2010) *We band of brothers: a true Australian adventure story*. Kindle version. Brian W McFarlane.

McSorley, K. (ed.) (2013) *War and the body: militarisation, practice and experience*. New York: Routledge.

Mitchell, B. (1989) *Weak link: the feminization of the American military*. Washington, DC; New York, NY: Regnery Gateway; Distributed to the trade by Kampmann.

Morgan, D. (1994) Theatre of war: combat, the military, and masculinities. In Brod, H. and Kaufman, M. (eds) *Theorizing masculinities*. London: Sage Publications. pp. 165–182.

Napier, G. (2011a) *A to Z of the Knights Templar a guide to their history and legacy* [online]. Stroud, Gloucestershire: History Press. Available at: http://public.eblib.com/choice/publicfullrecord.aspx?p=897139 (accessed: 24 July 2015).

Napier, G. (2011b) *The rise and fall of the Knights Templar* [online]. New York: The History Press. Available at: http://public.eblib.com/choice/publicfullrecord.aspx?p=897137 (accessed: 24 July 2015).

Norris, R. (2008) Mourning rights: *Beowulf*, the *Iliad*, and the war in Iraq. *Journal of Narrative Theory*, 37 (2): 276–295.

Peterson, V.S. (2010) International/global political economy. In Shepherd, L.J. (ed.) *Gender matters in global politics*. New York: Routledge. pp. 204–218.

Peterson, V.S. (2014a) Family matters: how queering the intimate queers the international. *International Studies Review*, 16 (4): 604–608.

Peterson, V.S. (2014b) Sex matters: a queer history of hierarchies. *International Feminist Journal of Politics*, 16 (3): 389–409.

Peterson, V.S. and Runyan, A.S. (2010) *Global gender issues in the new millennium. Dilemmas in world politics*. 3rd edn. Boulder, CO: Westview Press.

Peterson, V.S. and True, J. (1998) 'New Times' and New Conversations. In Zalewski, M. and Parpart, J.L. (eds) *The 'Man question' in International Relations*. Boulder, CO: Westview Press. pp. 14–27.

Pettman, J. (1996) *Worlding women a feminist international politics* [online]. London; New York: Routledge. Available at: http://public.eblib.com/choice/publicfullrecord. aspx?p=254158 (accessed: 7 November 2014).

Pin-Fat, V. and Stern, M. (2005) The scripting of Private Jessica Lynch: biopolitics, gender, and the 'feminization' of the U.S. military. *Alternatives: Global, Local, Political*, 30 (1): 25–53.

Pressfield, S. (2000) *An epic novel of the battle of Thermopylae*. [Place of publication not identified]: Bantam Books.

Pressfield, S. (2011) *The warrior ethos*. Los Angeles: Black Irish Entertainment.

Scaglione, A.D. (1991) *Knights at court: courtliness, chivalry & courtesy from Ottonian Germany to the Italian Renaissance*. Berkeley, CA: University of California Press.

Scheipers, S. and Greiner, B. (eds) (2014) *Heroism and the changing character of war: toward post-heroic warfare?* Houndmills: Palgrave Macmillan.

Shakespeare, William (1598) *Henry V*.

Shepherd, L.J. (2006) Veiled references: constructions of gender in the Bush administration discourse on the attacks on Afghanistan post-9/11. *International Feminist Journal of Politics*, 8 (1): 19–41. doi: 10.1080/14616740500415425.

Sjoberg, L. (2007) *Mothers, monsters, whores: women's violence in global politics*. New York: Zed Books.

Sjoberg, L. (2010) Gendering the empire's soldiers: gender ideologies, the US military, and the 'War on Terror'. In Sjoberg, L., Via, S., and Enloe, C. (eds) *Gender, war and militarism: feminist perspectives*. Santa Barbara, CA: Praeger. pp. 209–218.

Sjoberg, L. and Via, S. (eds) (2010) *Gender, war, and militarism: feminist perspectives*. Praeger security international. Santa Barbara, CA: Praeger.

Spivak, G.C. (1988) Can the subaltern speak? In Nelson, C. and Lawrence, G. (eds) *Marxism and the interpretation of culture*. Basingstoke: Macmillan Education.

Stephenson, M. (2012) *The last full measure: how soldiers die in battle*. 1st edn. New York: Crown Publishers.

Stern, M. and Zalewski, M. (2009) Feminist fatigue(s): reflections on feminism and familiar fables of militarisation. *Review of International Studies*, 35 (03): 611.

Sylvester, C. (2013) *War as experience: contributions from international relations and feminist analysis. War, politics and experience*. Milton Park, Abingdon, Oxon; New York, NY: Routledge.

Tickner, J.A. (1992) *Gender in international relations: feminist perspectives on achieving global security*. New York: Columbia University Press.

Urban, M. (2014) *The tank war: the British band of brothers – one tank regiment's World War II*. London: Hachette.

Van Creveld, M. (2000) A woman's place: reflections on the origins of violence. *Social Research*, 67 (3): 825.

Van Creveld, M. (2001) *Men, women, and war*. London: Cassell & Co.

Van Creveld, M. (2013) To wreck a military. *Small Wars Journal*. Available at: http:// smallwarsjournal.com/jrnl/art/to-wreck-a-military (accessed: 13 October 2015).

Weber, C. (2016) *Queer international relations: sovereignty, sexuality and the will to knowledge*. Oxford studies in gender and international relations. New York: Oxford University Press.

Welland, J. (2013) Militarised violences, basic training, and the myths of asexuality and discipline. *Review of International Studies*, 39 (04): 881–902.

Whitworth, S. (2007) *Men, militarism, and UN peacekeeping: a gendered analysis*. Critical security studies. Boulder, CO: Lynne Rienner Publ.

Winters, R.D. and Kingseed, C.C. (2011) *Beyond band of brothers*. London: Ebury.

Young, I.M. (2003) The logic of masculinist protection: reflections on the current security state. *Signs*, 29 (1): 1–25.

4 Grim reapers – narratives of masculinity and killing

Introduction

The ghost hunt of the warrior has revealed the numerous puzzle pieces that come together to create the spectral shape of the warrior. As such, the warrior is built out of various nuances, details, and particularities that are difficult to capture and this chapter foregrounds these complexities in relation to the lives of British Reaper crews. Perhaps the most contentious element of the lives of Reaper crews, and certainly the element over which most ink is spilt, is their capacity to use lethal force. Similarly, one of the core capabilities of the warrior is his capacity to inflict death on others and die gloriously himself. Through a ghost hunt of the experiences of Reaper crews in the use of lethal force, this chapter explores the densities of life that are the lived expression of the statement that 'life's complicated' (Gordon, 2008, p. 1).

Through a ghost hunt of this element of Reaper crews' experiences, this chapter draws attention to the inherent tension in narratives about killing, masculinity and military technologies. The capacity to kill is illuminated as a core marker of performing military masculinity, and a means through which the crews can access the warrior identity.[1] However, this is complicated by the Reaper crews' distance from the theatre of war and (relative) lack of physical risk, which has led to concerns about feminisation and cowardice.[2] Including 'sensuous' data makes it possible to explore the ways in which killing makes people *feel* uncomfortable and *at the same time* sediments the status of those who do; makes it possible to understand how killing can be interpreted through the language of *saving* lives; makes it possible to understand how political cultures that shy away from death can *simultaneously* fetishise those who die heroically and the social, cultural and historical weight woven through these narratives.

In this chapter I engage in conversation with two ghosts who emerge through the discussion of Reaper crews undertaking lethal strikes. These ghosts, as outlined in Chapter 3, are conceptual metaphors that do theoretical work – illuminating the interaction between textual and material forms of knowledge. Therefore, the ghostly figures of this chapter straddle the 'slash' between life and death in novel ways, explored through this chapter through the question of whether crews have a 'duty to kill?' and a 'duty to die?'. Through this I reflect on the way in which complex personhood can be acknowledged by presuming that life and people's lives are simultaneously straightforward and full of enormously subtle meaning' (Gordon, 2008, p. 5) in ways that have profound implications for the work that gender does.

The first ghost of the chapter is that of 'the Other', which for many feminist scholars, is an identity marker that signals the dangerous, the deviant, the different.[3] To this theorising I add the 'Other' of our own psychology, the subjectivity that comes from 'the life of others, and the other things within us' (Gordon, 2008, p. 48).[4] Following this, I use the ghost of the 'Other' to signify the way the crews interact with the targets of lethal strikes, the ways that they make sense of taking life, and the way that this is woven into narratives of activity, heroism, and paternal masculinity. This chapter argues that the masculinity of the Reaper crews is reinforced by the capacity to kill whilst *at the same time* destabilised (and therefore, perhaps, feminised) by the inevitable moments when the crews fail to *save* lives or cause the death of civilians. The (accidental?) death of civilians, particularly children, forces us to question the benign, parental masculinity that the crews establish through saving lives and disrupts discourses of the (moral) 'goodness' of the warrior figure more generally.

The second ghost of the chapter is that of cowardice, who acts as the 'unhappy echo' of the warrior attribute of courage (Woodward and Winter, 2007, p. 74). Through the prism of this ghost I ask whether warriors have a 'duty to die' a heroic death in order to sediment their masculine status and reflect on how this requirement is navigated in the (relatively) riskless roles of the Reaper crews. The requirement for members of the British military to act bravely is rendered 'uncanny' by the political aversion to casualties, creating an impossible environment where courage is simultaneously celebrated and undesirable. The chapter also explores the ghost of cowardice by questioning whether Reaper crews are as 'protected' and 'risk free' in their work and lives as they are generally portrayed.

Wielding the scythe

> When you sign up you know what you are signing up for … there is always the possibility you'll need to use lethal force … if you have a problem with that go work at the florist.…[5]

Being able to use 'kinetic effect', that is the ability to drop bombs, is an important part of establishing the usefulness and the 'warrior' status of drone crews.[6] But, as this section goes on to illustrate, killing is an act that is difficult to undertake, which raises questions without answers, and which requires engaging with complex ghostly others. Members of the British military have the right and responsibility to take life (under certain conditions) in pursuit of British military aims and within militaries those individuals who have permission to kill are awarded the highest status.[7] As outlined in Chapter 3, killing evokes the spectral mythology of physical strength (even where that is not directly relevant to the means of killing), the shared spilling of blood from which the Brother-at-Arms iteration materialises, and establishes the power status of the individual who can take life as dominant (and therefore masculine) in comparison to those who can be killed (who are feminised).

Whilst Carol Cohn and Sara Ruddick note that 'The practice of war entails far more than the killing and destruction of armed combat' there can be no question that killing and destruction are key identifying components of warfare (2003, p. 4).[8] But killing, even within the context of war, is difficult (and surely it should be).[9] Indeed

> the average and healthy individual ... has such an inner and usually unreal-
> ized resistance towards killing a fellow man that he will not of his own voli-
> tion take life if it is possible to turn away from that responsibility.
>
> (Marshall, 2000, p. 79)

The taboo against taking life is incredibly powerful, and troops are rendered uncomfortable by the training required to kill and by their own resistance to that training.[10] It is perhaps as a result of the energy expended and the difficulty associated with overcoming barriers to the use of lethal force that those who are able to do so, who do so 'well' (i.e. repeatedly and in the appropriate contexts) are given the higher status of warrior.

In addition to being troubled by the concurrent difficulty/discomfort with the act of killing *and* the higher status awarded to those who can and do kill, Reaper crews are situated within the gendered dualisms that connect masculinity with having the power to kill. The 'masculinity' of killing is discussed in a range of feminist works.[11] For example, Carol Cohn points to 'men's desire to appropriate from women the power of giving life ... conflat[ing] creation and destruction' (1987, p. 699). The status of destroyer/creator is one that is reflected in various iterations of military masculinity.[12] As Bourke claims 'for men, combat was the male equivalent of childbirth' because it was an 'initiation into the power of life and death' (Bourke, 1999, p. 14).[13] The connection between military masculinity and birth is also apparent in the lives of soldiers and aircrews. Bourke refers to 'officers being described as "giving birth" to a battery whilst also being exhorted to "father" their men' (1999, p. 145). In the lives of Reaper crews this phenomenon plays out in the relationship between the crews and the Reapers themselves. Matt Martin describes, in a manner surprisingly similar to that noted by Cohn (1987), how 'The Italians walked around the Predator "coffins", patting them as though they were the bellies of expectant mothers' (2010, p. 156). The fact that the transportation equipment for the Reapers were called "coffins" serves to highlight the connection between ideas of death and 'male parthenogenesis' (Braudy, 2005, p. 450).

Perhaps as a result of the connection between death and the desire for men to give birth to men (removing the feminine power to give life from women) the roles where the primary duty of the soldier is to 'close with the enemy' have traditionally been those to which women have been barred from applying (indeed, it was only in 2018 that that exclusion against women in close combat roles was lifted in the UK). The capacity to 'get a weapon off' (with all that statement's wonderful sexual undertones), an interviewee claimed, was important to attempting to resolve the manning crisis that the US Air Force was experiencing with Reaper.[14] Historically, during the Second World War the status of fighter pilots was equal to the number of 'kills' that they had under their belt with fighter pilots who were 'high scorers' becoming national heroes (Bourke, 1999, p. 48).[15] Therefore the missiles attached to Reaper, and the capacity to deploy them, are an important identifier of the crews' permission and capacity to kill, which in turn endows them with the masculine status of the warrior.[16]

This initial exploration reveals the complex way in which killing is a component part of what it means to be a 'man' in war. The narratives of the warrior are haunted by the need to kill others to maintain the masculine identity and therefore the successful execution (in both senses of the word) of the duty was/is to be celebrated. The capacity to kill in war is then understood as the capacity to do one's duty, if it is acknowledged that killing is a key component of what makes war. Therefore being permitted to take life, to drop bombs, and engage the enemy is important not just because it enables the individuals to feel like active parts of the military machine, but because it reflects masculinity woven into the fabric of the British forces.[17] However, as the next section draws out, the capacity to kill and the connection with masculinity is not straightforward, and in the lives of British Reaper crews, this connection is troubled by the ghost of the 'Other'.

The ghost of the 'Other'

The complexity of personhood can be seen in the lives of British Reaper crews. Empowered by the experience of being *able* to kill they are troubled by the times that they kill, the times that they do not kill and the times that they might have to kill:

> We struck the checkpoint; the missile went exactly where it was supposed to. It wasn't a 'normal' feeling, it was an abnormal thing to do. Even talking about it now I feel my stomach turning. It was gut-wrenching that I've now taken people's lives. It's a weird feeling.
>
> (Interview with a Reaper operator in Lee, 2018, p. 210)

The warrior status, the technological capabilities, the press perspectives and personal/professional moral compasses all weave together to make for a situation that is haunted by the complexity of navigating competing claims and identities of this role. This section begins by exploring the interaction between the warrior status and technological capabilities, before moving on to illuminating how this interaction is complicated by the ghost of 'the Other', situating the crews at the centre of the complex web of components listed above.

As part of the British military, Reaper crews gain social status from crewing aeroplanes that have weapons and who can claim to have deployed those weapons. Flight Lieutenant Garrick Hill (RAF) notes that

> The original Predator UAVs were principally used in the reconnaissance role but the MQ-9 Reaper is a hunter-killer UAV.... They're equipped with Hellfire missiles and Paveway laser-guided bombs, which is pretty much the same pay load as one of the USAF's F-16s.
>
> (Loveless, 2010, p. 195)

This similarity with fast jet weapons (often considered the 'top' of the hierarchy of military masculinity) is important as Dan noted, 'I was quite often in the US Air Force bar when the cream of the crop of Air Force pilots [would come in], they thought they were the bee's knees, but had never ever been in a combat situation ...' and therefore, he could establish his status by asking them, 'How many bombs have you dropped? None? Oh I've done five this week'.[18] The pride that crew members feel at being operationally busy is enhanced by the fact that they can do more than *just* provide surveillance, countering some criticisms of Reaper squadrons from within the RAF. Whilst the fast jet guys were still perceived as wrinkling their noses in distain, other crews 'had a really operational focus' and the Reaper crews could say 'we're operational all the time, and we're dropping bombs', which enabled them to settle more comfortably into the social hierarchy.[19] Therefore, the 'pride associated with flying RPAs' appears to be largely based around 'getting the most combat experience of anyone' (Bergen and Rowland, 2013, p. 113).

The ghost of the 'Other' draws attention to the complexity in these webs of gender which can be better understood by reflecting on how complex personhood is being lived in these contexts. The warrior status that confers the marker of military masculinity on the Reaper crews is one that must be constantly performed and renegotiated. Through this lens, the criticism of the Reaper crews results from the way in which their lethal capacity stabilises their own gendered status but the capacity to kill at such an extreme distance (and the lack of physical risk to the crews), erodes the mythic power of the dog fight, the historical legacy on which the fighter pilots draw their status.

The ghost of the 'Other' also gestures to the Reaper crews' complex relationship with friendly forces on the ground. How can an apparently riskless crew perform warrior masculinity in comparison to those brave souls risking their lives below? Almost to a man my interviewees stated 'saving lives' was the best part of the job.[20] 'We're not killing people for the fun of it. It would be the same if we were the guys on the ground. You have to get to [the enemy] somehow or all of you will die' (Reaper crew member cited in Woods, 2015). The importance of saving lives was reflected in the introduction of the unofficial motto (for 39 Squadron) of, 'To save lives. To make a difference'. Therefore, having bombs is important not just because it establishes their credential as warriors but because it allows them to save lives. Saving the lives of fellow soldiers is particularly important because the persistent surveillance of some of the ground force overwatch missions meant that the crews feel like they get to know the guys on the ground (and even the enemy – as is explored in Chapter 5).

Providing 'overwatch' enables drone crews, who are not at physical risk, to undertake the role of protector and thereby reclaim some of the traditional attributes of military masculinity.[21] Temporality is disturbed by the ghost of the 'Other', because the different time-zones in which the crews operate enable the Reaper to provide overwatch for sleeping troops on the ground. In an interview, 'Tim' 'recalls driving home to Las Vegas, realising that he had achieved something special for a large group of people' by providing overwatch:

> The Comms Officer on the ground apologised for bringing out the Reaper when nothing had happened. And then he added: 'Just to let you know, I am the only person who is awake on this FOB. Everyone else is getting some sleep and it's the first sleep they've had in days.
>
> (Lee, 2018, p. 80)

Perhaps similarly, Robert noted, his fondest memory was providing overwatch for some Canadian soldiers so that they could get a good night's sleep without fear of being ambushed.[22] Where the lives of the individuals on the ground are at risk, the drone crews provide masculine protection through watching, and if necessary, deploying lethal force.

The lens of in/(hyper)visibility illuminates the way in which the crews could come to 'recognize the faces and figures of our soldiers and marines, unbeknownst to most of them' in ways that emotionally affected the watching crews: 'I sometimes chuckled over their youthful pranks and high jinks … I cried with them as well whenever they lost a comrade' (Martin, 2010, p. 121). Similarly, one interviewee told me that it 'broke [his] heart' watching the 'young lads' in the infantry 'being sick [with fear] before they leave' the camp; language which speaks of the emotional connection between the Reaper crews and the ground troops ('We took it personally'), and reveals a paternal perspective.[23] The language here seems to reflect the idea of the Reaper crews as benign, almost parental figures, rolling their eyes at the mischief that the 'kids' on the ground get up to. Therefore, it is unsurprising that the most painful iteration of the ghost of the 'Other' are the moments when the Reaper crews are unable to save lives and troops die: 'If anybody on the Reaper fleet says it doesn't affect them, then they're lying. It does. It has to' (Lee, 2018, p. 299).

The complexity of the personhood of Reaper crews is illuminated by this interaction between killing and saving lives, by the masculinity of killing and the masculinity of *preventing* death. Whilst the 'fog of war' means that the death of friendly forces/comrades is a virtual inevitability in all conflicts, this section addresses how this reality plays out in the lives of British Reaper crews (Clausewitz *et al.*, 1993). Reaper crews are constantly on operations, they work long shifts, and have huge amounts of information to assimilate and sift through. It is, therefore, understandable that on occasions they are not able to/fail to prevent the death of friendly forces, of 'lads on the ground'. The connection between the Reaper crews, and the troops on the ground was outlined in the previous section, and given this connection, situations in which those individuals suffer injury or death that could or might have been prevented by Reaper crews creates a sensation of helplessness and emasculation: 'That flash on the screen, and the feeling impotence, just stayed with us. Our job was to provide overwatch on those guys, to protect them' (Lee, 2018, p. 75). In an interview with Adrian Chiles, former Wing Commander of 13 Squadron Damian Killeen reflects on the impact of these kinds of experiences:

> KILLEEN: I've had this conversation with mates of mine as well, trying to relate it to them, and erm … the ones that have watched the film 'American Sniper', there's a bit in that where he turns around and says 'It's not the engagements I've made that bother me, it's the ones that I haven't made that bother me'…. The bit that hurts more are the days where you watch people die because we are in a surveillance mode. You know for example, you'll be in a surveillance mode and the guys you're protecting stand on an improvised explosive device, or you are watching….
> CHILES: But there's nothing you could have done about that
> KILLEEN: There's nothing you could have done about it but….
>
> (Killeen, 2015)

Being in surveillance mode (being unarmed) or having permission to strike withdrawn by the chain of command can increase the sense of impotence, the feeling of emasculation. Having watched a scenario, described as 'carnage' unfold after having permission to strike withdrawn, 'Gav' asked 'Boss can you find out why clearance for the shot was withdrawn. We would like to know why we are watching friendlies on fire', his barely suppressed rage shuddering through the controlled language, particularly as the crew were then tasked to go back to the same area so that 'from the infrared images on the screen they could see the heat draining from the bodies of the dead Iraqi soldiers' (Lee, 2018, pp. 188, 191).

It is perhaps partly as a result of this and the battle damage assessments (BDA) that must be carried out after missile strikes that make crewing Reapers traumatic. One of the reasons that my interviewees (and those interviewed by others) are so hostile to accusations that they are emotionally distanced from their operations, is because what they see (particularly where this is combined with a sense of helplessness) can be so distressing. BDA forms an important part of the tasking of Reaper crews, and this first section outlines some of the ways in which this trauma manifests (Asaro, 2013; Woods, 2015). Given the nature of the task it is not surprising that crew members report seeing distressing things as part of this role:

> When you hit a truck full of people, there are limbs and legs everywhere … I watched a guy crawl away from the wreckage after one shot with no lower body. He slowly died. You have to watch that. You don't get to turn away.
>
> (Maurer, 2015)

The smoke clears, and there's pieces of the two guys around the crater. And there's this guy over here, and he's missing his right leg above his knee. He's holding it, and he's rolling around, and the blood is squirting out of his leg, and it's hitting the ground, and it's hot. His blood is hot. But when it hits the ground, it starts to cool off; the pool cools fast. It took him a long time to die. I just watched him. I watched him become the same colour as the ground he was lying on.

(Bryant cited in Power, 2013)

I know the feeling you experience when you see someone die. Horrifying barely covers it. And when you are exposed to it over and over again it becomes like a small video, embedded in your head, forever on repeat, causing psychological pain and suffering that many people will hopefully never experience.

(Linebaugh, 2013)

Whilst these quotes are from articles that deal with the experiences of US drone crews, interviewee data reveals similar experiences for British Reaper crews, although there was a reluctance to speak in too much detail: 'You watch him the entire time you are striking … [this is] the emotional aspect of killing'.[24] If you saw friendly forces struck it was sometimes difficult to get back to the task at hand: 'our job is not to watch the vehicle, watch the number of chest compressions he was given … you have to come back out to make sure that they are safe …' stated Geoff, whilst his statement clearly indicated that the crews had *wanted* to count the chest compressions, to see if the guy made it. As Peter noted, having a 'front row of watching own soldiers getting blown up' is sometimes 'a mentally traumatising experience'.

Experiencing war as an emotionally traumatising experience has previously been attributed to men who were effeminate – those individuals who deviated from the stoical male warrior norm. But this myth has endured. None of the individuals I interviewed claimed to have suffered psychological injury (perhaps because of this legacy of emotional stoicism), but a number referred to others that they knew who had. Drone Wars UK Chris Cole's interview with 'Justin Thompson' makes similar claims:

JT: I've seen one or two videos which I wish I hadn't watched, even though I'm used to combat and had the dubious honour of having to watch combat up very, very close …

CC: And doing battle damage assessment after launching a strike I guess …

JT: Exactly and that is not pleasant either. It's better in IR [infra-red] than in day TV, I can tell you that. And yes, I felt a good deal of the kind of stress that that would induce in any normal human being. I'm not ashamed to admit I have shed tears from time to time, but I think I coped with it in a fairly healthy way. I was quite good at compartmentalising and I was also quite good at letting it come to the fore when it needed to. I know of people who have suffered quite badly with PTSD.

(Cole, 2017)

And since I started this book more and more Reaper crews have come forward to speak about emotional distress. Interviewee Peter claimed that 'I know that some people have suffered PTSD … the fact that you are not physically there is countered by ability to zoom in … to some fairly horrible sights'. Lee's book includes interviews with an individual who sought medical help but whose distress was not officially classified as PTSD, with the doctor noting 'you have elements of it. But you don't have PTSD because you're not re-living it'. Acting as important reminder that emotional distress, and psychological injury manifests in a range of different ways, so that the interviewee notes 'Mine is more of a traumatic memory, I don't have enough ticks in the boxes for it to be full PTSD. I was on tablets – antidepressants – for two or three months' indicating that even when not classed as 'full' PTSD the results can be debilitating and require medical and psychological interventions (Lee, 2018, p. 296).

What then, is the implication for the crews who observe these kinds of things? The commentary above reminds us of the importance of situating these narratives within the framework of Haunting as that provides for an understanding of the complexity of personhood. After all, the statements above suggest that whilst there are some individuals and some occasions (people after all do not always respond to the same stimuli in the same ways on different days) who feel distanced from the 'reality' of what occurs on the ground and who are able to place a buffer between themselves and the things that they see, for some individuals this separation is not possible. The reference to the effects of adrenaline on crews' bodies – the heart palpitations etc. draws on the capacity for Haunting to incorporate both the discursive and the material, the different ways that 'sensuous knowledges' can be explored (Holloway and Kneale, 2008). In the past emotional responses to the trauma of war have been labelled as a 'lack of moral fibre', a failure of masculinity, a failure to be a man in the face of danger. Whilst the British military certainly takes its duty towards securing the mental health of its forces seriously, there remain serious barriers to acknowledging the emotional impact of war, including for Reaper crews.

Abort! Abort! Abort!

Given the social and cultural importance attached to the capacity to kill within the British military and the desire to protect friendly forces on the ground, Reaper crews are, at times, under considerable pressure to release weapons. However, it is important to note that in some cases *not* releasing weapons can be constructed as not only the most appropriate, but also potentially heroic way of proceeding in an operation. The desire to protect friendly forces must be carefully balanced against the risk to civilians or the 'combatant' status of individuals who may or may not pose a threat to ground troops. The ghost of the 'Other' highlights this tension between the two facets of Reaper crew roles. Given that they are not at physical risk in the same way, their awareness of the difference between their situation and the individuals on the ground raises the spectre of the deaths of 'Others' – making it difficult for them to refuse to act. This balancing act of protecting ground troops and preventing the deaths of 'innocents' brings us to the next sections on courageous restraint and accidental deaths.

The complexity of navigating these different facets of masculinity is partly a result of the legal and ethical structures put in place to restrict the situations in which British Reaper crews operate (laws of war, codes of conduct for the British Armed Forces, rules of engagement for specific operations) (Armed Forces Act, 2006; House of Commons Defence Committee, 2013). Additionally, there have been recent novel social/cultural developments that have foregrounded *not* killing, *not* striking as appropriate behaviour for military professionals. These developments have been framed around the concept of 'courageous restraint'. Introduced by General McChrystal during operations in Afghanistan, courageous restraint 'restricted the rules of engagement for British and US forces so that they could then only fire back if they clearly identified their target, rather than putting down fire and risking civilian lives' (Dixon, 2012, p. 40). For Reaper crews this means being prepared to resist commands (and requests) to take action where they are concerned that it might result in civilian/friendly forces deaths. As such, crew members operate at the nexus, in a liminal zone, between the masculine glorification of killing, the benign patriarchal stance of saving lives, and resisting the demands to do either. This is particularly difficult to do when there are men on the ground at risk, and when those requesting support have to be denied it. One individual reported a response of 'Brothers are going to die because of you', which 'cut us to the core' even though knew they needed to ensure that the 'wrong' people did not die (Lee, 2018, p. 284). An example of how courageous restraint works for Reaper crews can be found in an RAF Operational Update, which outlines the following:

> The Kandahar-based Reapers of 39 Squadron were engaged in operations that again necessitated Hellfire strikes this week. Displaying considerable tactical restraint, the Reaper crews successfully countered the insurgents' best efforts to seek cover among civilians and along tree lines.
>
> (S, 2011)

And another example

> Whilst tracking an insurgent on a motorbike carrying a suspicious package the crew of the Reaper suspected that the 'package' may be a small child. Showing outstanding courageous restraint, the crew used replay facilities to provide sufficient evidence to delay and subsequently terminate the attack. When the motorbike finally stopped at a residential compound the action of the crew was fully justified when a small child got off. In a separate attack a

single Hellfire was used to great effect on a group of insurgents which resulted in seven enemy forces being killed in action.

(S, 2011)

During discussions interviewees were keen to emphasise the way in which they were empowered to refuse to strike. This capacity to say 'no' to undertaking strikes was an important theme in discussion and interviewees described a 'red card system', which refers to the right and the responsibility to call a halt to a planned strike if they have concerns about meeting the criteria of the ROE, about endangering life, or for any other reason.

One of the reasons that having the capacity to refuse to strike is so important is because of the trauma and torment that civilian and/or friendly deaths cause to the crews. Whilst the ghost of the 'Other' certainly haunts the lives of British Reaper crews through the capacity and the military necessity of the act of killing, this ghost has a second role. Reflecting the core theme of in/(hyper)visibility, the ghost of the 'Other' also plays out in the fear of killing the wrong individuals and of collateral damage. One interviewee described a 'quite horrible scenario' where friendly forces were under fire and he was preparing to 'go live' with weapons when 'a variable [said] don't do it' and on closer inspection a 'flutter' turned out to be a 'group of kids'.[25] The 'variable' spoken of was the murkiness of the imagery, a hunch that something *something* was not what it appeared so that in this instance a reliance on sensuous knowledge and intuition saved lives.

Sometimes crews do cause civilian deaths. Two incidents have been described, one in 2011 in Afghanistan and one in 2018 in Syria. In the former, a strike against two Taliban vehicles resulted in the deaths of several women and children who were passengers, unbeknownst to the crews (or apparently intelligence analysts) (Lee, 2018). In the latter, confirmed by the MOD, a civilian motorcyclist was caught in the blast against an ISIS vehicle (Cole, 2018). One of the individuals involved in the 2011 incident recalls 'Two questions dominated: What is that child doing there? And why didn't we know about it?', as a result the individual struggled to come to terms with what had happened:

How did I find myself in this situation? I joined the Reaper Force to make a positive difference after the shit I experienced in Afghanistan.... What did I do? ... What could I have done differently? ... We took innocent lives.

(Lee, 2018, pp. 102, 107)

Unsurprisingly, he had to 'think long and hard about whether I wanted to get back into the [Ground Control Station] box again' (Lee, 2018, p. 110) Undertaking strikes in populated areas, even with precision technology, is always dangerous and sometimes comes with horrifying consequences. These consequences often 'haunt' the Reaper crews, creating trauma, flashbacks and question marks about choices made or not made.

The ghost of cowardice[26]

> War is about many things, but at its core it is about killing or getting killed.
>
> (Stephenson, 2012, p. 355)

The previous section of this chapter has looked at the importance of killing, this section now turns to the opposite side of the coin, that is 'getting killed'. Through the ghost of cowardice this section investigates the way that contemporary risk aversion is disturbed by the historical and mythological legacy of what it *means* to be brave in war. In this section time and space are disrupted as narratives of the way military technology can and should be used to protect 'our soldiers' collide with discourses of risk and reciprocity in the fetishised battle form of the duel.

Western militaries are investing in new technologies that distance warriors from the theatre of war because dying in war is no longer politically acceptable.[27] Whereas Napoleon 'boasted of his army's capacity to tolerate deaths in battle' (Smith, 2005, p. 487), maintaining support for an extended military campaign in many of today's democracies (the UK included) depends at least in part in the ability to pursue military objectives without sustaining high numbers of deaths to personnel: 'one is not "allowed" to get killed' (Kober, 2015, p. 97). As a result, there is a tension between the political perspective and the cultural and social context of the British armed forces, one which reflects the queer logic(s) of the ghost of cowardice. Baggiarini notes 'the deaths of citizen-soldiers are (inconsistently) profane' (2015, p. 130). They are 'inconsistently' profane because we continue to fetishise the heroic warrior who dies in war whilst allowing those deaths to erode support for military operations. The contradiction built into the social and political context of the British armed forces is illustrated by quotes from my interviewees:

> Most honours and awards are given, rightly so, to soldiers directly in combat or harm's way for their response to enemy fire and when aiding their comrades or the civil population, when their own lives are on the line.[28]

> [within the military leadership there is a] massive amount of risk aversion.[29]

> we don't want to see people in body bags.[30]

Comparing these quotes, it would appear that we want 'our boys' to come back in one piece, not only as a moral imperative, but because it is important for maintaining political support for operations: 'we don't want to see people in body bags'.[31] However, those 'safe' individuals are less likely to be promoted and are less likely to be acknowledged for their efforts either through promotion boards or in the process of selecting individuals to receive medals.[32]

Commentators have directly connected the increase in risk aversion in the armed forces with their degradation (or, in the words of Martin Van Creveld, 'feminisation').[33] The ethical and cultural importance of reciprocal risk between combatants has been remarkably resistant to change (despite technological innovation), reflecting the power of the myth of duel, even in the contemporary western military.[34] In the RAF specifically the heroic status of fighter pilots was cemented by their 'duels' or dog fights with German (or other) fighter pilots during the World Wars.[35] Whilst it has been argued extensively that drones are no more than the logical extension of technology that has steadily distanced the individuals involved in war, there remains a disquiet over the role of distance in enabling killing and in ideas of valour and gallantry.[36] Indeed, some claim that 'the drone is a cowardly weapon that threatens to make cowards of those who embrace it' (Plaw and Fricker, 2015, p. 198), so apparently the most honourable way to return from battle still reflects the dictum of the Spartan mothers: 'Come back with your shield, or on it' (Fields, 2013, pp. 64–65), reminding us that 'All the good men are already dead' because they demonstrated their courage by performing their 'duty to die' in war (Braudy, 2005, p. 6).

The importance of risk to creating the warrior–pilot identity was described by interviewee Roxy who noted that the recent boom in promoting helicopter pilots is at least in part because 'these are the guys who have been winning the war, getting shot at and winning the crosses'.[37] 'Winning the crosses' as a recognition of the bravery of these pilots is important for institutional respect and has resulted in a high number of helicopter crew members being promoted to positions of power. In previous iterations of this process, other 'risky' roles were over represented in the higher echelons of the RAF, described by the same individual as the 'Harrier mafia' and 'Typhoon mafia'. Similarly, the risk inherent in flying and the status that came with that role is recognised through the provision of 'flying pay' to flight active personnel, which was traditionally perceived as 'danger money' (Sims, 1967). Interviewee Noel argued that the status awarded to individuals with the 'riskier' roles was informally as well as formally recognised, describing the hierarchy in the bars on bases: 'are you air crew or not, pilot or not, are you multi, rotary or fast jet (with RPAS as fourth pillar), fast jet tornado or single seat, which squadron' with the highest status being conferred upon those who could lay claim to flying the best (read: riskiest) aircraft.[38]

If warrior behaviour requires an individual to place his or herself at risk and then to act courageously, it then follows that cowardice may be considered the very antithesis of the warrior identity. Cowardice has specific meaning in military-legal terms. The 1955 Army Act defines cowardly behaviour as

> Any person subject to military law who when before the enemy – (a) leaves the post, position or other place where it is his duty to be, or (b) throws away his arms, ammunition or tools, in such a manner as to show cowardice....
>
> (Army Act, Chapter 18, p. 18)

As such, the definition is a relatively narrow one and during the First World War 306 men were executed for cowardice.[39] More recent military legal documents

(Armed Forces Act, 2006; House of Commons Defence Committee, 2013) no longer include specific definitions of, or in general reference to 'cowardice' acknowledging that amongst those executed for cowardice in times of war were many who were likely contentious objectors and individuals suffering from PTSD.[40] However, this is not to imply that cowardice does not continue to have *cultural* salience both within the British military and in commentary *about* the British military. In the following section I outline these different and important ways of understanding the contemporary meaning and usage of cowardice and how it relates to Reaper crews.

Chris Walsh's seminal book on the topic indicates that 'mentions of *cowardice* have risen over the last decade, and the uptick for the term *coward* is sharper still' suggesting that the term is being deployed in new and important ways (2014, p. 9). Whilst the updated Army Act (2006) does not refer to cowardice as an offence, the phrase is used in the UK Armed Forces Personnel and Legal Framework for Future Operations (2013). In this document the abuse of Baha Mousa, who died in British custody in Iraq in 2003, is described as 'violent and cowardly' (House of Commons Defence Committee, 2013, p. 25). Similarly, in the documentation of the investigation into the abuse, one of the abusers admits that 'he was guilty of an unprovoked and cowardly assault' on Baha Mousa (Gage, 2011, p. 249). Cowardice, it would appear, has morphed from an incapacity for violence to something rather different. In this case, the term describes the inhumane actions of individuals in a position of power over another who could not fight back. It is in this sense that the epithet of coward is (problematically) attached to drone operations.

Drone operations are described by some as inherently cowardly. In an interview with *Vice* Abu Mosa of the jihadist group ISIS/Daesh told the US 'Don't be cowards and attack us with drones. Instead send your soldiers, the ones we humiliated in Iraq' (quoted in Johnson, 2014; see also Monbiot, 2012). Other similar perspectives include from 'Mobashar Jawed Akbar, founding editor of the Asian Age and a former senior fellow with the Project on U.S. Relations with the Islamic World at the Brookings Institution' who 'argues that the American use of drones will be interpreted as an act of cowardice, not strength' and Glenn Greenwald who states that

> Whatever one thinks of the justifiability of drone attacks … [attacking by drone is] one of the least 'brave' or courageous modes of warfare ever invented…. Killing while sheltering yourself from all risk is the definitional opposite of bravery.
>
> (Both cited in Goldstein, 2015, pp. 72, 73)

In the same piece Cora Sol Goldstein notes that counterinsurgency expert David Killcullen 'contends that "using robots from the air … looks both cowardly and weak"' (quoted in Goldstein, 2015, p. 73). Whilst I am certainly not claiming that this perspective is accurate, these statements are instructive in understanding the cultural and political context in which Reaper crews operate.

As such, even taking into account that cowardice is sometimes wrongly seen in behaviour that is, rather, 'really prudent or even courageous', 'coward' remains a powerfully unpleasant epithet (Walsh, 2014, pp. 10, 11). The question is whether it is socially preferable to undertake physical risk in warfare rather than 'risk' being called a coward: 'the shame of cowardice reinforces the call of sacrifice', hence my question about a duty to die.[41] For example, Kaurin states

> To be branded a coward is amongst, if not, the worst insults that can be levelled at a warrior, and it is not merely an indictment of his individual character, but also a criticism of his commitment to his cause and his fellow warriors at a very deep and existential level.
>
> (2014, p. 13)

In connecting cowardice with unmanliness Paul Robinson claims that 'Most men will have internalized the idea that courage is a virtue, and cowardice a sin, and will wish to avoid the sense of shame that arises from failing to live up to what they believe are society's expectations' (2009, p. 4).[42] Similarly, Braudy describes the relationship between cowardice and the idea of failed masculinity stating that 'the word for cowardice in Greek is *anandareia*, literally "unmanliness"' (2005, pp. 32–33).[43]

Is it the case then, as Cara Daggett claims, that '[b]ecause drone operators are protected from death, they are disqualified from performing as 'real' warriors because their bodies are not sited in combat' (2015, 363) reflecting the fear that the way Reaper crews are protected from physical harm somehow renders them morally suspect and unmanly (Royakkers and van Est, 2010; Asaro, 2013)? Given the connection between risk, courage and military masculinity, it is no surprise that being stationed at an extreme distance from the theatre of war has a negative impact on the way the Reaper crews are viewed by their colleagues and by the press: 'he flies model aeroplanes out of Vegas ... he's not ... at risk'.[44] Interviewee Ken stated that there are some 'disparaging misconceptions', a sense that Reaper crews were somehow 'war dodging ... by not committing their bodies'. If you are not sharing the risk then you are stalked by the spectre of cowardice, and 'the lack of risk affects how you are perceived within the military'.[45] If you are not sharing the risk, then you are not acting as a warrior and you are stalked by the spectre of cowardice.[46] To be heroic, to meet the requirements for warrior masculinity, the Reaper crews need to do something that risks their bodies, in a way that their roles are not considered to enable them to do.

The historical need for warriors to prove their worth by risking their bodies continues to haunt contemporary militaries despite changes to the technological and political components of warfare. The impact of the sense that that you are 'not as brave' as the guys in the fast jets, the helicopters, the people on the ground with 'their balls hanging out', can be seen in the recent debate over medals for Reaper crews in the United States (Allsep, 2013, p. 389).[47] An expression of the perception that 'from an aviators point of view, if you don't cross the border you don't get a medal' and that '[w]hat matters most to [other members of the military] is "being there" in battle, and all they see are the drone pilots' distant duty stations' (Spangler, 2013). Whilst not quite labelling Reaper crews as cowards, these statements and other similar ones clearly connect risklessness in war with this image. As Lee notes, Reaper crews challenge 'historical tradition[s] for awarding medals for geographically specific campaigns and the associated "risk and rigour" in theatre' (2018, p. 230). The debate about whether or not Reaper crews should be given medals for service was, perhaps unsurprisingly, a heated one but in July 2018 'the UK Defense Secretary announced that the Operational Service Medal Iraq and Syria medal "will now recognise.... Reaper pilots taking life and death decisions"' (Lee, 2018, p. 249). The ghost of cowardice acts as a challenge to the construction of Reaper crews as warriors, and therefore to their ability to perform the military masculinity constructed as an integral part of identity in the British military.

The Reaper crews themselves indicate a sense, a hunch, that being a warrior means, or requires, that physical personal risk. For example, the response to the question: 'Is an RPAS crew member a warrior? Please describe why/why not' elicited a range of responses from agreement with gusto, to the kind of self-effacement that is frequently ascribed to 'gallant' British military individuals, to an (obviously emotional and uncomfortable) complete rejection of the concept in any contemporary military context. Responses included:

[An] RPAS crew member is not a warrior, that's the guys on the ground.

You need warrior culture [in RPAS crews].

RPAS are warriors, in that they are an elitist entity that specialises in warfare using the weapons available to them, to the best of their ability.

There is no concept of heroism or warrior-ship in the military. All members will act to save themselves and others – that is not heroic – that is our job.

Absolutely, he is an unrecognised warrior.

I've never seen myself as a warrior … it's the wrong word, I don't think you can be a warrior if you're not there.[48]

As such, interviews with former Reaper crew members and other members of the RAF revealed a complex relationship between the figure of the warrior and the current cultural requirements of a modern professional military. It became clear through these discussions that the low level of physical risk[49] to Reaper crews and their relationship with those individuals on the ground whose bodies *were* at risk of injury or death was one central point around which conceptualisations of what constituted warrior behaviour coalesced.

Rethinking narratives of risklessness

The mythological hero worship of bodies at risk and death in warfare has created a complex environment in which Reaper crews negotiate their subjectivity given their comparative lack of physical risk in their experience of war. However, whilst the preceding sections have outlined the *implications* of the claim that 'Physically, drone pilots are so safe, in fact, that descriptions of drone piloting at Nevada's Creech Air Force Base ... sound downright monotonous' this section challenges the *substance* of the claims (Beauchamp, 2016). As such, I argue that the assertion that Reaper crews live *completely* riskless combat lives is not as accurate as it first appears.

Where the previous section explored the temporal dislocation between our war stories and the technological and political reality of war on the basis of experiencing personal risk, this section brings the attention back to the complexity of personhood as the crews are *perceived* as functioning in a riskless environment when parts of their roles are not as risk free as is popularly believed. Claims about the lack of physical risk to Reaper crews are based on the existence of a separate and distinctive battlefield, one that is different from the 'home front' (a division explored in greater detail in Chapter 6). Situating the individual on that battlefield is then something that endangers them and renders them masculine, whereas being based 'at home' is connected with safety and femininity. However, this distinctive and separate theatre of war is something that is becoming increasingly challenged in contemporary warfare. Not only are Reaper crews operating from their home or third countries, but combatants are killing and maiming combatants and civilians in areas that are traditionally considered to represent 'the home' or areas of non-violence.[50] Instead of functioning on clearly marked battlefields Reaper crews (and other members of the British Armed Forces) are increasingly based in grey, shadowy spaces between war and peace, between home and abroad, dislocating themselves geographically, temporally and, perhaps, legally and ethically. Therefore as Blair notes, 'I do not believe that RPA operators are in less danger than their manned counterparts.... This war is global, and our enemies have global reach as well' (Blair, 2012, p. 63). Blair's assertion is corroborated by interview data that indicates two sources of risk in the lives of British Reaper crews. The first is the risk that crews are under when they are deployed to the Forward Operating Bases (FOBs) for launch and recovery of the Reapers and the second is the risk that they face in their 'home' base from terrorists and other individuals opposed to the use of drones.[51]

The risk posed to the Reaper crews at the FOB is simple. Deployed 'in theatre' these crews are tasked with the take-off and landing parts of Reaper missions, which required 'line of sight'[52] control and therefore cannot be done from the extreme distance of the main missions.[53] Being deployed in theatre means that these crews are at the same degree of risk as any other forces deployed to the same FOB and that the crews' lives are overlaid with the history of those troops who have gone before. In his memoir, US Reaper pilot Martin noted that

> I was both nervous and excited … I would be, more or less, among the rear-echelon types that front-line grunts called 'Fobbits' – those who remained at the forward operating base (FOB) while everyone else went out on patrol. Nonetheless, this would be my first time actually setting foot in a war zone.
>
> (Martin, 2010, p. 140)

Situated at the FOB, in the *space* of the war, Reaper crews are at risk, and plagued by the possibility of death and injury from incoming artillery, bombing, missiles and sniper attacks. Martin recalls, 'A few nights later, Balad sounded *Alarm Red* – base under attack' as missiles were fired into the compound (2010, p. 201). This happened so frequently that Martin responded to the firework celebrations for 4 July by throwing himself to the floor and waiting for the 'next shell to crash through the roof' (2010, p. 202). Whilst the fireworks represented a shock rather than an actual threat, the preceding quotes indicate that it is clearly not the case that Reaper crews are *never* at risk of physical harm in their roles. That Martin responded in this way indicates that the shock of being under attack haunted him even after the event, with the adrenal system of his body responding to the phantom of death from missiles and bombs.

The second form of risk that Reaper crews experience is risk of attacks that take place away from the theatre of war. That terrorists are inherently phantasmal and intrude into our lives in unexpected ways is something already explored within the spectralities literature.[54] In the context of Reaper operations, as Blair noted the 'war' (or 'operations' as titled by the UK forces) is global, and the 'enemy' does not respect the boundaries of the theatre of war.[55] Indeed, acting as combatants, even from such an extreme distance, means that Reaper crews can be understood as legitimate targets for attack, including on US and UK soil. The legality of targeting Reaper crews was acknowledged by US General McChrystal who stated that 'Anything we use that's part of that [defence infrastructure] becomes, in my view, a fair target for our enemies, and we have got to consider that' (quoted in Norton-Taylor and Ross, 2015). This sentiment was echoed in the UK by Mr David Davis (MP of Haltemprice and Howden) who noted:

> If we undertake drone attacks outside a defined war zone, the location from which those attacks are operated may become part of a war zone, and we may legitimise a counter-attack on that area. Since many of those places are in rear areas, [and] that has real issues for the operation of our RAF bases, such as RAF Waddington.
>
> (Column 50WH, Hansard, 2015)

The implication is that an attack on Waddington or Creech would be considered a legitimate/legal strike on British or American military leadership (Norton-Taylor and Ross, 2015). These statements demonstrate that whilst military technology might have changed the way that British aircrews operate, the 'enemy' has also changed the way s/he operates in response, negating some of the gains towards risklessness.

The impact of this perceived risk of strikes on bases and on the targeting of Reaper crew members became visible through my interviews and fieldwork. Whilst crews did not want to be perceived as fearful, interviewees spoke of a need for caution in discussing their roles and revealing their identities to the wider public. As one interviewee baldly stated, 'I am in no way shape or form ashamed of what I do but I cannot say what I do because of the need to protect my family'.[56] This sentiment was corroborated by interviewees who claimed that threats had been made against personnel and one interviewee recalled being visited by the British Counter-Terrorism Police unit who had found his name on a list of targets of a suspected terrorist cell in the UK. The crews are haunted by the murder of Lee Rigby, a ghost who acts as a reminder of the crews' fleshy and destructible bodies as they travel to and from the base, acknowledging that, 'You are a target on the road to Creech'.[57] As a result the crews' roles and identities as Reaper pilots/sensors/analysts become phantasmal – 'nobody in the village knows what they do' in an attempt to protect themselves and their families from violence.[58] My rationale for anonymising my interviewees was not so much as a measure to enable them to speak freely, but rather an important means of protecting them from harm, from becoming targets for terrorist operations. Therefore, we know the crews exist but we cannot know who they are.

Therefore, despite claims that Reaper crews are not at risk, it appears that there are two significant ways in which they are at physical risk of death and injury. Given the increasing use of terrorist tactics, such as suicide bombings, and the perception that Reaper crews are 'fair game' as targets, their lives are not as sheltered and secure as many commentators might believe.[59] The masculinity of the crews, destabilised by their distance from the battlefield can be (re)inscribed through the assumption of physical risk from such attacks. Perhaps more than any other component, questions of risk and the warrior identity demonstrate the way in which the lives and experiences of British Reaper crews simultaneously destabilise and (re)inscribe our traditional understandings of military masculinity.

Conclusion

Killing is a core component of the identity of the warrior, and a primary marker of what it means to engage in warfare. Implementing the framework of Haunting in this chapter I have drawn attention to the way that Reaper crews' lives are complicated by their capacity to undertake lethal strikes. The heart of this chapter is the desire to take complex personhood seriously, to explore the densities of the various narratives at work that influences the crews' experiences of killing. The masculine marker of killing is set uncomfortably against narratives of saving lives, which haunt the background of operations where unintended killing occurs – either of friendly forces by enemies or of 'innocents'. The masculinity of having the capacity to kill is situated in apparent opposition to the (relative) lack of risk to the lives of the crews themselves; the technology representing the culmination of the political (and strategic) desire to protect one's own troops.

However, what I have drawn attention to in this chapter, through an engagement with 'sensuous knowledges' and my addition to the framework of queer logic(s), is the way apparent oppositions in binaries can co-exist. The capacity to kill can empower *at the same time* as disempower, the way someone *feels* can impact the way that they *think* about the experience of killing, and the way that they construct their identity is woven out of what they *thought* being a warrior *was* and how they act in their day-to-day life.

Engaging with 'sensuous knowledges' materialised the two ghosts of this chapter: the ghost of the 'Other' and the ghost of cowardice. The 'Otherness' of the first ghost reflects feminist theorists' concerns with the use of the 'Other' as a marker of dangerous differences that has often been applied to women.[60] To this Haunting adds the concern of the 'Other' within, reflecting its psychoanalytical background.[61] This ghostly 'Other' then is the ghost of the experience of killing. Engaging this ghost in conversation I ask whether Reaper crews have a *duty* to kill, and what it means for these individuals when they make sense of killing as *saving* lives. Constructed this way it is possible to shine a light on one of the contradictions at the heart of this complex experience and to consider the particular ramifications for subjectivity (in both senses of the word) when lives cannot be saved, when the 'wrong' people die.

The second ghost of the chapter, cowardice, I engaged in conversation by asking whether, as a counterpoint to killing, the crews have a *duty to die*. A controversial statement, I use this argument to frame the discussion of the continuing salience of sacrifice for the warrior identity and the implications related to the reduction of risk. One of the defining features of the warrior, as outlined in the previous chapter, is his bravery and capacity to continue in the face of severe threats to his personal safety: to life and limb. Because, as many commentators would have it, the Reaper crews are not at risk, their status as 'protected' renders them both feminine and cowardly, I introduced the description of the two ways in which the crews *are* physically at risk and how this serves to highlight the ways in which, in this kind of conflict in particular, things are not as straightforward as they might initially seem. I used this data to engage with the ghost of cowardice, illuminating the way that this ghosts illuminates the queer logic(s) at work in the *simultaneous* political aversion to casualties *and* the continuing power of narratives of heroic death.

This chapter has focused on killing and death in the lives of British Reaper crews, which is certainly the most contentious part of their roles. However, it is a small component of their work, most of which involves conducting persistent surveillance. Continuing the ghost hunt, in the following chapter I move on to consider how the experience of watching and being watched is an integral part of destabilising and (re)inscribing the crews' warrior identities.

Notes

1 (Cohn, 1987; Goldstein, 2001).
2 (Asaro, 2013; Royakkers and van Est, 2010; Daggett, 2015).

3 (Richter-Montpetit, 2007; Sjoberg, 2007; Pettman, 1996; Hansen, 2006).
4 (after Freud, 1997; Abraham and Torok, 1999; Abraham *et al.*, 2008).
5 Interview with 'David'.
6 This was particularly important to those who had not used a weaponised platform before (Interview with 'Steve').
7 (Bourke, 1999).
8 As Shane Riza notes 'In war, killing must be done', a statement borne out by the (cob) web of deaths which connects the ghostly warriors sketched out in the previous chapter (2013, p. 35; see also Grossman, 2009; Bourke, 1999; Wright, 1965).
9 (Goldstein, 2001).
10 (Bourke, 1999; Riza, 2013).
11 (see for example Downing, 2013; Peterson and Runyan, 2010; Bourke, 1999).
12 (Cooke, 1996; Hooper, 1998; Morgan, 1994).
13 (see also Broyles Jr., 1984; Schott and Heinämaa, 2010).
14 Interview with 'Geoff' (see, for example, Alexander, 2015; Drew and Philipps, 2015; Majumdar, 2015; Chatterjee, 2015; Terkel, 2014; Siminski, 2013).
15 (see also Hillary, 2010; Sims, 1967).
16 (Cohn, 1987; Grossman, 2009; Higate, 2003).
17 (Woodward, 2004; Woodward and Duncanson, 2016; Francis, 2008; Higate, 2003).
18 Interview with 'Dan'.
19 Interview with 'Tom'.
20 I had only one female interviewee and she did not comment on this issue.
21 (Pin-Fat and Stern, 2005; Elshtain, 1995; Hicks Stiehm, 1982).
22 Interview with 'Robert'.
23 Interview with anonymous intelligence analyst.
24 Interview with 'Dan'.
25 Interview with 'Robert'.
26 I recognise that including the phrase 'a duty to die' is controversial, indeed members of the military can be disciplined for taking unnecessary risks (author's discussion with David H. Dunn). However, as feminist scholars have indicated in a range of different ways, those military personnel who lose their lives in the pursuit of their objectives (who put themselves as risk of death) are awarded the highest social status within their groups and British society more widely (Baggiarini, 2015; Millar and Tidy, 2017; Robinson, 2009).
27 (Luttwak, 1995; Coker, 2002; Calhoun, 2011).
28 Interview with 'Ben'.
29 Interview with 'James'.
30 Interview with 'Roxy'.
31 Interview with 'Roxy'.
32 (see Blair, 2012, p. 66 for discussion).
33 (Coker, 2007; Van Creveld, 2013; Kober, 2015).
34 (Bourke, 1999; Lee, 2012).
35 (Rosenberg, 1993; Francis, 2008; Braudy, 2005, p. 55).
36 (Beauchamp, 2016; Schulzke, 2014; Enemark, 2011; Alston and Shamsi, 2010; Royakkers and van Est, 2010).
37 Interview with 'Roxy'.
38 Interview with 'Noel'.
39 (Sweeney, 1999).
40 (Olsthoorn, 2007; Inbar *et al.*, 1989; Walsh, 2014; Sweeney, 1999).
41 (Walsh, 2014, p. 12; see also Robinson, 2009).
42 I would argue that this is something not *just* experienced by men but because of the connection between masculinity, physical strength and war it has particular salience to this discussion.
43 (see also Robinson, 2009; Ehrenreich, 1998; Bourke, 1999).

44 Interview with 'Roxy'.
45 Interview with 'Ken' and 'Peter'.
46 As one interviewee noted 'the RPAS tour was initially viewed as the equivalent of a rest tour', with some individuals being sent there in the posting equivalent of a 'pat on the back', because Creech was viewed as a posting on the strip in Vegas where you flew a model aeroplane as you sipped cold cocktails by the pool in your back yard. (Allsep, 2013, p. 389; Blair, 2012, p. 63; Beauchamp, 2016).
47 Interview with 'Ken' and 'Tom'.
48 Interviews with 'James', 'Noel', 'Steve', 'Ian', 'Roxy', and 'Amy'.
49 Despite claims to the contrary (Coeckelbergh, 2013; Royakkers and van Est, 2010), Reaper crews are not shielded from all possibility of physical violence and danger associated with warfare, as will be outlined later in the chapter.
50 For example, terrorist strikes on shopping malls and schools, or Reapers dropping bombs on populated areas and homes where targets are meeting.
51 (for example Worley, 2016; Pawlyk, 2016).
52 'Line of sight' refers to the need to be able to visually see the Reaper craft during take-off and landing rather than being able to operate it 'over the horizon' via satellite link (Department of Defense Report, 2012).
53 (Loveless, 2010, p. 196; Martin, 2010, p. 160).
54 (Engle, 2009; Auchter, 2014).
55 As interviewee 'Robert' claimed (with problematic 'othering'): 'these people don't respect borders'.
56 Interview with 'Anonymous'.
57 Off-duty Fusilier Lee Rigby was murdered in 2013 in broad daylight by Islamic extremists. Interview with 'Freddie' and 'Amy'.
58 Interview with 'Robert'. The need for some level of secrecy is emphasised by recent postings by individuals purporting to be members of ISIS/Daesh who released a list of 72 names that they claim belong to US Reaper crews with the instruction that their followers 'Kill them wherever they are, knock on their doors and behead them, stab them, shoot them in the face or bomb them' (cited in Worley, 2016).
59 (for example Vallor, 2013; Asaro, 2013; Cole *et al.*, 2010).
60 (Sjoberg and Via, 2010; Peterson and Runyan, 2010).
61 (Gordon, 2008; Abraham and Torok, 1999; Abraham *et al.*, 2008; Freud, 1997).

References

Abraham, N. and Torok, M. (1999) *Cryptonymie: le verbier de l'homme aux loups*. Champs 425. Paris: Flammarion.

Abraham, N., Torok, M. and Rand, N. (2008) *L'écorce et le noyau*. Paris: Flammarion.

Alexander, D. (2015) Air Force moves to ease stress on overworked US drone pilots. *Reuters*, 16 January. Available at: http://in.reuters.com/article/2015/01/16/usa-defense-drones-idINL1N0UU35M20150116 (accessed: 7 March 2015).

Allsep, L.M. (2013) The myth of the warrior: martial masculinity and the end of don't ask, don't tell. *Journal of Homosexuality*, 60 (2–3): 381–400. doi: 10.1080/00918369.2013.744928.

Alston, P. and Shamsi, H. (2010) A killer above the law? *Guardian* [online], 8 February. Available at: www.theguardian.com/commentisfree/2010/feb/08/afghanistan-drones-defence-killing (accessed: 4 January 2015).

Armed Forces Act (2006). Available at: www.legislation.gov.uk/ukpga/2006/52/pdfs/ukpga_20060052_en.pdf (accessed: 3 December 2017).

Army Act, Chapter 18. Legislation: Army Act of 1955. Available at: www.legislation.gov.uk/ukpga/1955/18/pdfs/ukpga_19550018_en.pdf

Asaro, P.M. (2013) The labor of surveillance and bureaucratized killing: new subjectivities of military drone operators. *Social Semiotics*, 23 (2): 196–224. doi: 10.1080/10350330.2013.777591.

Auchter, J. (2014) *The politics of haunting and memory in international relations. Interventions.* London; New York: Routledge/Taylor & Francis Group.

Baggiarini, B. (2015) Drone warfare and the limits of sacrifice. *Journal of International Political Theory*, 11 (1): 128–144. doi: 10.1177/1755088214555597.

Beauchamp, S. (2016) Can drone pilots be heroes. *The Atlantic*, 23 January. Available at: www.theatlantic.com/politics/archive/2016/01/can-drone-pilots-be-heroes/424830/ (accessed: 27 January 2016).

Bergen, P. and Rowland, J. (2013) Drone wars. *The Washington Quarterly*, 36 (3): 7–26. doi: 10.1080/0163660X.2013.825547.

Blair, D. (2012) Ten thousand feet and ten thousand miles: reconciling our air force culture to remotely piloted aircraft and the new nature of aerial combat. *Air & Space Power Journal*, pp. 61–69.

Bourke, J. (1999) *An intimate history of killing: face-to-face killing in twentieth-century warfare.* New York: Basic Books.

Braudy, L. (2005) *From chivalry to terrorism: war and the changing nature of masculinity.* 1. Vintage Books ed. New York: Vintage Books.

Broyles Jr., W. (1984) Why men love war. *Esquire*, November.

Calhoun, L. (2011) The end of military virtue. *Peace Review*, 23 (3): 377–386.

Chatterjee, P. (2015) Our drone war burnout. *New York Times*, 14 July. Available at: www.nytimes.com/2015/07/14/opinion/our-drone-war-burnout.html?_r=0 (accessed: 22 October 2015).

Clausewitz, C. von, Howard, M., and Paret, P. (1993) *On war.* London: David Campbell.

Coeckelbergh, M. (2013) Drones, information technology, and distance: mapping the moral epistemology of remote fighting. *Ethics and Information Technology*, 15 (2): 87–98. doi: 10.1007/s10676-013-9313-6.

Cohn, C. (1987) Sex and death in the rational world of defense intellectuals. *Signs*, 12 (4): 687–718.

Cohn, C. and Ruddick, S. (2003) *A feminist ethical perspective on weapons of mass destruction.* Consortium on Gender, Security and Human Rights Working Paper No. 104. Available at: http://genderandsecurity.org/sites/default/files/carol_cohn_and_sara_ruddick_working_paper_104.pdf (accessed: 19 September 2014).

Coker, C. (2002) *Waging war without warriors?: the changing culture of military conflict.* IISS studies in international security. Boulder, CO: Lynne Rienner Publishers.

Coker, C. (2007) *The warrior ethos: military culture and the war on terror.* LSE international studies series. London; New York: Routledge.

Cole, C. (2017) Interview of former RAF Reaper pilot 'Justin Thompson' (a pseudonym) by Chris Cole, *Drone Wars UK*, May 2017.

Cole, C. (2018) UK drone strike kills civilian in Syria admits MoD. *Drone Wars UK.* Available at: https://dronewars.net/2018/05/02/uk-drone-strike-kills-civilian-in-syria-admits-mod/ (accessed: 20 December 2018).

Cole, C., Dobbing, M. and Hailwood, A. (2010) Convenient killing: armed drones and the 'Playstation' mentality. [online]. Available at: http://dronewarsuk.files.wordpress.com/2010/10/conv-killing-final.pdf

Cooke, M. (1996) *Women and the war story.* Berkeley, CA: University of California Press.

Daggett, C. (2015) Drone disorientations: how 'unmanned' weapons queer the experience of killing in war. *International Feminist Journal of Politics*, 17 (3): 361–379.

Department of Defense (2012) MQ-9 Reaper unmanned aircraft system (MQ-9 Reaper), RCS: DD-A&T(Q&A)823-424.

Dixon, Paul (2012) *The British approach to counterinsurgency: from Malaya and Northern Ireland to Iraq and Afghanistan.* London: Palgrave Macmillan.

Downing, L. (2013) *The subject of murder: gender, exceptionality, and the modern killer.* Chicago, IL: The University of Chicago Press.

Drew, C. and Philipps, D. (2015) As stress drives off drone operators, air force must cut flights. *New York Times*, 16 June. Available at: www.nytimes.com/2015/06/17/us/as-stress-drives-off-drone-operators-air-force-must-cut-flights.html?ref=us&_r=0 (accessed: 29 June 2015).

Ehrenreich, B. (1998) *Blood rites*. St Ives, UK: Virago.

Elshtain, J.B. (1995) *Women and war.* University of Chicago Press ed. Chicago, IL: University of Chicago Press.

Enemark, C. (2011) Drones over Pakistan: secrecy, ethics, and counterinsurgency. *Asian Security*, 7 (3): 218–237.

Engle, K. (2009) *Seeing ghosts: 9/11 and the visual imagination.* Montreal: McGill-Queen's University Press.

Fields, N. (2013) *The Spartan way.* Barnsley, South Yorkshire: Pen & Sword Military.

Francis, M. (2008) The flyer: British culture and the Royal Air Force, 1939–1945. Oxford; New York: Oxford University Press.

Freud, S. (1997) The uncanny. In *Writings on art and literature.* Stanford: Stanford University Press. pp. 193–233.

Gage, W. (2011) *The Baha Mousa Public Inquiry report – Volume I.* HC 1452–1. House of Commons. Available at: www.gov.uk/government/uploads/system/uploads/attach ment_data/file/279190/1452_i.pdf (accessed: 3 December 2017).

Goldstein, J.S. (2001) *War and gender: how gender shapes the war system and vice versa.* Cambridge: Cambridge University Press.

Goldstein, C.S. (2015) Drones, honor, and war. *Military Review*, 95 (6): 70.

Gordon, A. (2008) *Ghostly matters: haunting and the sociological imagination.* New University of Minnesota Press ed. Minneapolis, MN: University of Minnesota Press.

Grossman, D. (2009) *On killing: the psychological cost of learning to kill in war and society.* Rev. ed. New York: Little, Brown and Co.

Hansen, L. (2006) *Security as practice: discourse analysis and the Bosnian war. The new international relations.* New York: Routledge.

Hansard (2015b) Drones in conflict [online]. House of Commons. Available at: www. publications.parliament.uk/pa/cm201516/cmhansrd/cm151013/debtext/151013-0004. htm#151013-0004.htm_spnew20

Hicks Stiehm, J. (1982) The protected, the protector, the defender. *Women's Studies International Forum*, 5 (3/4): 367–376.

Higate, P. (ed.) (2003) *Military masculinities: identity and the state.* Westport, CT: Praeger.

Hillary, R. (2010) *The last enemy.* London: Vintage.

Hooper, C. (1998) Multiple masculinities in international relations. In Zalewski, M. (ed.) *The 'man question' in international relations.* Boulder, CO: Westview Press.

Holloway, J. and Kneale, J. (2008) Locating haunting: a ghost-hunter's guide. *Cultural Geographies*, 15 (3): 297–312. doi: 10.1177/1474474008091329.

House of Commons Defence Committee (2013) *UK armed forces personnel and the legal framework for future operations.* HC931. House of Commons. Available at: www.

publications.parliament.uk/pa/cm201314/cmselect/cmdfence/931/931.pdf (accessed: 3 December 2017).

Inbar, D., Solomon, Z., Spiro, S. and Aviram, U. (1989) Commanders' attitudes toward the nature, causality, and severity of combat stress reaction. *Military Psychology*, 1 (4): 215–233.

Johnson, A. (2014) ISIS to US: 'Don't be cowards and attack us with drones ... send your soldiers'. *National Review*. Available at: www.nationalreview.com/corner/384981/isis-us-dont-be-cowards-and-attack-us-drones-send-your-soldiers-andrew-johnson.

Kaurin, P.M. (2014) *The warrior, military ethics and contemporary warfare: Achilles goes asymmetrical*. Military and defence ethics. Farnham, Surrey: Ashgate Publishing Company.

Killeen, D. (2015) *BBC Five Live Adrian Chiles interviews UK drone pilots*. Available at: www.bbc.co.uk/programmes/b060xx5w#auto (accessed: 7 September 2015).

Kober, A. (2015) From heroic to post-heroic warfare: Israel's way of war in asymmetrical conflicts. *Armed Forces & Society*, 41 (1): 96–122. doi: 10.1177/0095327X13498224.

Lee, P. (2012) Remoteness, risk and aircrew ethos. *Air Power Review*, 15 (1): 1–20.

Lee, P. (2018) *Reaper force: the inside story of Britain's drone wars*. London: John Blake Publishing Ltd.

Linebaugh, H. (2013) I worked on the US drone program. The public should know what really goes on. *Guardian*, 29 December. Available at: www.theguardian.com/commentisfree/2013/dec/29/drones-us-military (accessed: 3 May 2016).

Loveless, A. (2010) *Blue sky warriors: the RAF in Afghanistan in their own words*. Sparkford, Yeovil, Somerset, UK: Haynes Publishing.

Luttwak, E. (1995) Toward post heroic warfare. *Foreign Affairs*, 74 (3): 109.

Majumdar, D. (2015) US drone fleet at 'breaking point', Air Force says. *The Daily Beast*, 1 May. Available at: www.thedailybeast.com/articles/2015/01/04/exclusive-u-s-drone-fleet-at-breaking-point-air-force-says.html (accessed: 16 March 2016).

Marshall, S.L.A. (2000) *Men against fire: the problem of battle command*. Norman, OK: University of Oklahoma Press.

Martin, M.J. (2010) *Predator: the remote-control air war over Iraq and Afghanistan: a pilot's story*. Minneapolis, MN: Zenith Press.

Maurer, K. (2015) She kills people from 7,850 miles away. *The Daily Beast*, 18 October. Available at: www.thedailybeast.com/articles/2015/10/18/she-kills-people-from-7-850-miles-away.html (accessed: 2 February 2016).

Millar, K.M. and Tidy, J. (2017) Combat as a moving target: masculinities, the heroic soldier myth, and normative martial violence. *Critical Military Studies*, 3 (2): 142–160. doi: 10.1080/23337486.2017.1302556.

Monbiot, G. (2012) With its deadly drones, the US is fighting a coward's war. *Guardian*, 30 January. Available at: www.theguardian.com/commentisfree/2012/jan/30/deadly-drones-us-cowards-war (accessed: 3 December 2017).

Morgan, D. (1994) Theatre of war: combat, the military, and masculinities. In Brod, H. and Kaufman, M. (eds) *Theorizing masculinities*. London: Sage Publications. pp. 165–182.

Norton-Taylor, R. and Ross, A. (2015) RAF base may be legitimate target for Isis, says ex-Nato commander. *Guardian*, 25 November. Available at: www.theguardian.com/uk-news/2015/nov/25/raf-base-may-be-legitimate-target-isis-ex-nato-commander (accessed: 31 March 2016).

Olsthoorn, P. (2007) Courage in the military: physical and moral. *Journal of Military Ethics*, 6 (4): 270–279.

Pawlyk, O. (2016) ISIS-linked hackers claim to release personal information of U.S. drone pilots. *Air Force Times* [online], 5 March. Available at: www.airforcetimes.com/story/military/2016/05/02/isis-linked-hackers-claim-release-personal-information-us-drone-pilots/83852686/ (accessed: 17 May 2016).

Peterson, V.S. and Runyan, A.S. (2010) *Global gender issues in the new millennium. Dilemmas in world politics.* 3rd edn. Boulder, CO: Westview Press.

Pettman, J. (1996) *Worlding women a feminist international politics* [online]. London; New York: Routledge. Available at: http://public.eblib.com/choice/publicfullrecord.aspx?p=254158 (accessed: 7 November 2014).

Pin-Fat, V. and Stern, M. (2005) The scripting of Private Jessica Lynch: biopolitics, gender, and the 'feminization' of the U.S. military. *Alternatives: Global, Local, Political*, 30 (1): 25–53.

Plaw, A. and Fricker, M.S. (2015) *The drone debate: a primer on the U.S. use of unmanned aircraft outside of conventional theatres of war.* Lanham, MD: Rowman & Littlefield.

Power, M. (2013) Confessions of a drone pilot. *GQ*, 23 October. Available at: www.gq.com/news-politics/big-issues/201311/drone-uav-pilot-assassination (accessed: 18 September 2014).

Richter-Montpetit, M. (2007) Empire, desire and violence: a queer transnational feminist reading of the prisoner 'abuse' in Abu Ghraib and the question of 'gender equality'. *International Feminist Journal of Politics*, 9 (1): 38–59.

Riza, M.S. (2013) *Killing without heart: limits on robotic warfare in an age of persistent conflict.* 1st edn. Washington, DC: Potomac Books.

Robinson, P. (2009) *Military honour and the conduct of war.* Routledge Military Studies. Abingdon: Routledge.

Rosenberg, S.D. (1993) The threshold of thrill: life stories in the skies over Southeast Asia. In Cooke, M.G. and Woollacott, A. (eds) *Gendering war talk.* Princeton University Press. pp. 43–66.

Royakkers, L. and van Est, R. (2010) The cubicle warrior: the marionette of digitalized warfare. *Ethics and Information Technology*, 12 (3): 289–296. doi: 10.1007/s10676-010-9240-8.

S, A. (2011) *RAF operational update October 2011.* Available at: www.fox2.co.uk/viewtopic.php?t=13308&mode=print&sid=3c383a9c80c4cc240e5ffb656818cf70 (accessed: 4 July 2016).

Schott, R.M. and Heinämaa, S. (eds) (2010) *Birth, death, and femininity: philosophies of embodiment.* Bloomington: Indiana University Press.

Schulzke, M. (2014) The morality of remote warfare: against the asymmetry objection to remote weaponry: the morality of remote warfare. *Political Studies*, p. n/a–n/a.

Siminski, J. (2013) Nobody wants to fly drones. *The Aviationist*, 9 June. Available at: http://theaviationist.com/2013/09/06/nobody-wants-to-fly-drones/ (accessed: 16 March 2016).

Sims, E.H. (1967) *The fighter pilots: a comparative study of the Royal Air Force, the Luftwaffe and the United States Army Air Force in Europe and North Africa, 1939–1945.* London: Cassell.

Sjoberg, L. (2007) *Mothers, monsters, whores: women's violence in global politics.* New York: Zed Books

Sjoberg, L. and Via, S. (eds) (2010) *Gender, war, and militarism: feminist perspectives. Praeger security international.* Santa Barbara, CA: Praeger.

Smith, H. (2005) What costs will democracies bear? A review of popular theories of casualty aversion. *Armed Forces & Society*, 31 (4): 487–512. doi: 10.1177/0095327X0503100403.

Spangler, S. (2013) Being there: UAV crews & combat valour. *Jet Whine*, 25 April. Available at: www.jetwhine.com/2013/04/being-there-uav-crews-combat-valor/#sthash. lCvH6y3V.dpuf.

Stephenson, M. (2012) *The last full measure: how soldiers die in battle*. 1st edn. New York: Crown Publishers.

Sweeney, J. (1999) Lest we forget: the 306 'cowards'' we executed in the first world war. *Guardian* [online]. Available at: www.theguardian.com/world/1999/nov/14/firstworld war.uk (accessed: 3 December 2017).

Terkel, A. (2014) Drone pilots are suffering from low morale: GAO report. *Huffington Post*, 18 April. Available at: www.huffingtonpost.com/2014/04/18/drone-pilots_n_ 5173569.html (accessed: 16 March 2016).

Vallor, S. (2013) *The future of military virtue: autonomous systems and the moral deskilling of the military*. Tallinn: NATO CCD COE Publications.

Van Creveld, M. (2013) *Wargames: from gladiators to gigabytes*. Cambridge; New York: Cambridge University Press.

Walsh, C. (2014) *Cowardice: a brief history*. Princeton, NJ: Princeton University Press.

Woods, C. (2015) Drone warfare: life on the new frontline. *Guardian*, 24 February. Available at: www.theguardian.com/world/2015/feb/24/drone-warfare-life-on-the-new-frontline (accessed: 7 March 2015).

Woodward, R. (2004) Discourses of Gender in the Contemporary British Army. *Armed Forces & Society*, 30 (2): 279–301.

Woodward, R. and Winter, T. (2007) *Sexing the soldier: the politics of gender and the contemporary British Army. Transformations, thinking through feminism*. London; New York: Routledge.

Woodward, R. and Duncanson, C. (2016) Gendered divisions of military labour in the British armed forces. *Defence Studies*, 16 (3): 205–228.

Worley, W. (2016) Isis hacking group claims it has a mole in the Ministry of Defence. *Independent* [online], 5 February. Available at: www.independent.co.uk/news/uk/ home-news/isis-hacking-group-mole-in-ministry-of-defence-a7010011.html (accessed: 17 May 2016).

Wright, Q. (1965) *A study of war*. 2nd edn. Chicago, IL: University of Chicago Press.

5 The spectral screwdriver – on watching and being watched

It is very difficult to let go ... because you're always on ops ... omnipresent

(Robert)[1]

With 500 people looking over your shoulder, you daren't screw up.

(Dan)[2]

Introduction

One of the core concerns of Haunting is that of in/(hyper)visibility (see Chapter 2). This is partly because the word 'spectral' 'evoke[s] an etymological link to visibility and vision', both that which is *looked at* and that which is doing the *looking* (Blanco and Peeren, 2013, p. 2). It is also because visibility, seeing, and being seen are always tangled into intricate webs of power. *Who* can be seen, *how* they can be seen, how that *seeing* is interpreted are core questions raised in the implementation of the framework of Haunting. In this chapter, I trace in/(hyper)visibility through a ghost hunt of the queer logic(s) of British Reaper crews' experiences of persistent surveillance and of *simultaneously* being heavily scrutinised themselves. In the British context, Reaper crews are primarily used for Intelligence, Surveillance and Reconnaissance (ISR) and as many of my interviewees noted, 90–95 per cent of their tasking reflects this.[3] In addition to being tasked with watching, the Reaper crews themselves are closely watched by military leadership, politicians, and the popular press; described by one interviewee as 'the world's longest screwdriver'.[4] Drawing on forms of 'sensuous knowledge', as outlined in Chapter 2, I argue that the Reaper crews' experiences of watching are haunted by femininity, which contrasts with the masculinist ocularcentricity of traditional understandings of surveillance. Tracing these logics from the scientific Enlightenment period, I explore how claims to objectivity and rationality are built on the premise of seeing and being seen. Additionally, in keeping with the framework laid out in Chapter 2, this chapter asks what is it that the crews are *not* seeing? What is it that is *not* there? And what is the effect of (and indeed affect which results from) what *is* seen? Bearing in mind that the ghostly queerly crosses the boundary between the visible and invisible – they are traces, suggestions, and hints, this chapter focuses on how the ghostly muddles Enlightenment conceptualisations of reality and destabilise claims to gendered behaviour on the basis of that reality. Additionally, addressing the 'hyper' of the neologism in/(hyper)visibility, this chapter looks at how previously invisible crews are rendered 'other' through their sudden hypervisibility in sensationalist reporting – their activities, particularly their conduct of war (lethal strikes and surveillance) is rendered strange even as explaining that strangeness unravels.

To recap from Chapter 2 then, the ghostly disrupts what is considered visible and what is invisible (in the strict sense), and the boundary between these two types of data is challenged. The in/(hyper)visible refers to the way in which, although we tend to prioritise data that can been seen, there are also ways of 'knowing', or of becoming aware of something that is not seen or is not visible.[5] What is or is not visible is not just empirically important but has political implications – as noted in Chapter 2, what is visible but ignored, what is rendered hypervisible and therefore deemed abnormal/unusual/deviant/dangerous, are all woven tightly into webs of power, privilege that always also disempower and restrict. As Kyle Grayson and Jocelyn Mawdsley note, citing (Rose, 2007), 'vision always raises "questions of social difference, social relations and social power"' (2018, p. 12). Therefore, Haunting is useful for mapping the politics of in/(hyper)visibility *alongside* the other facets of complex personhood, disturbed temporality, and power.

Referring to the ghostly figure of the warrior of Chapter 3, this chapter explores how Reaper crews destabilise *and* (re)inscribe conceptions of military masculinity and femininity through their experiences of surveillance. In traditional gendered binaries the masculine is represented in the position of the watcher, and the feminine is relegated to the position of the object that is watched.[6] Therefore, this chapter argues that the experiences of British Reaper crews are an example of a situation where individuals can be construed as situated in the position of the masculine watcher *and* the feminine watched, at the same time as troubling the distinction between those two categories and exposing the limitations of binary thinking. The first part of the chapter briefly sketches out the importance of surveillance in warfare and of the technologies used to conduct it. From this I argue that one of the core concerns of Haunting, specifically in/(hyper)visibility is intertwined with the gendered logics of the watcher/watched dyad. Through this prism the chapter explores the ways in which masculinist ocularcentricity is both (re)inscribed and destabilised by the Reaper crews' recourse to subjective assessments and sensuous knowledges in making 'information' from 'data'.

The second section of the chapter, focuses on one of the bizarre elements of Reaper crew surveillance, which is the development of a curious 'distant intimacy'.[7] Utilising Haunting's psychoanalytic background mapped across the wider concern of in/(hyper)visibility, this section traces the way that what the crews *see*, what is made *visible* to them, is simultaneously comforting and traumatising.[8] Focusing on instances where the crews recognise a similarity between themselves and their targets ('even those called "Other" are never,

never that' (Gordon, 2008, p. 5)), I argue that Haunting provides the means through which to understand how these experiences and emotions can co-exist, and how they play into and challenge narratives of military masculinity.

The third part of the chapter flips the concerns of in/(hyper)visibility on its head by situating the crews, rather than their targets, as the feminised objects of scrutiny and speculation. Beginning from the perspective of the crews as 'missing' or 'invisible' in early interest in drone warfare I illustrate the ways in which the humans behind the machine were initially erased from the narratives about the technology that they use. From being invisible, shadowy beings, the Reaper crews then begin to materialise in newspaper articles and scholarly journals. However, in these instances the crews are rendered *hyper*visible – constructed as dangerous 'Others' who challenge norms of warfare and the identity of the warrior. Through the prism of Haunting I illuminate the ways in which popular fear of drones and the future of drone warfare infuses discourses of the crews who become feminised (as weakling armchair warriors) and infantilised in a double shot of public derision.

Continuing with the perspective of the crews as the feminised object, the final section of the chapter focuses on the scrutiny of the crews by their fellow members of the British Armed Forces. This section traces the way in which the connectivity of the Reaper enables military leadership to 'get back in the cockpit' with the crews, acting as a ghostly presence that erodes military professional trust.[9] Countering the argument that Reaper crews experience this as *only* a feminising experience I finish the chapter by arguing that this scrutiny *simultaneously* provides the necessary audience for the performance of military masculinity through technological mastery of an aircraft that is difficult to fly. What emerges throughout the chapter is the various ways in which gendered questions of in/(hyper)visibility are functioning in relation to Reaper operations and that it is only through engaging in a ghost hunt inflected with queer logic(s) that it is possible to draw out all these nuanced complexities.

An uncanny kind of surveillance

The importance of surveillance in war is nothing new, however the advent of drones has enabled a specific iteration: persistent surveillance from an extreme distance.[10] Predator (the precursor to Reaper) was originally designed as a reconnaissance asset and the capacity to launch missiles was only added later.[11] Whilst the missile strikes from drones are the focus of much of the debate about drones, it is surveillance that takes up the vast majority of the crews' time and energy, as one individual described: 'We spend 70 to 80 per cent of our time … just scanning roads' (Drew, 2009). Because there is no crew on board of Reaper 'it can spend some twenty-four hours in the air, flying at heights of up to twenty-six thousand feet' by simply swapping between crews at the ground-based controls, allowing US operator Martin to claim 'We were always present over the war front, watching, waiting' (2010, p. 29). My interview data describes a similar focus on surveillance as a core role and capability for British Reaper tasking:

Geoff noted that '90% of missions [were] surveillance' and David noted that these were important 'from a continuity perspective for detecting that pattern of life....'[12] This kind of long-term watching, interviewees argued, helps to save lives and prevent civilian casualties, for example: Geoff noted one near-miss where the crews

> spent 6 hours watching two guys dig by the side of the road ... [thinking] they must be implanting an IED ... we didn't fire a weapon ... they don't know we are watching them ... about a month later [after learning more about local behaviours] we re-watched it and could tell they were irrigating their field.[13]

Such operations encouraged 39 Squadron to adopt the unofficial motto of *Adjuvamus tuendo* – 'We assist by watching'.[14]

However, watching via drone is not a straightforward process of using an extended pair of binoculars. There is a complex web of electrical, digital, satellite communications that convey an astonishing amount of data, up to 6,000 terabytes per day, unless an object is being track in which case up to one exabyte of data (that's one million terabytes) can be produced from cameras with 'up to 65 vantage points' (Grayson and Mawdsley, 2018, p. 13). Whilst this appears to support ideas of the omniscience of Reaper operators it is not quite that simple. Not only are there some limitations to the technology – the oft-quoted line that Reaper vision can be like 'looking through a straw' but there are significant barriers to the human processing of all of this data, even with the assistance of various supercomputers (Lee, 2018, p. 43).

The creation of what Stahl (2013) refers to as 'drone vision' emerges from the complex range of imaging components in the 'nose' of the Reaper and provides something different from what is seen from the cockpits of fast jets. As an interviewee stated the idea of the 'video game mentality' that is frequently connected to the type of surveillance undertaken by Reaper crews was entirely back to front: when he described dropping bombs from a Tornado, 'Tom' noted that he would have to be so focused on 'setting up the tactical stuff', focused on a 'tiny screen and so many things to think about' (not least keeping the plane in the air) that he 'barely had a chance to look at the target' and was 'completely disconnected from the end result'.[15]

Reaper has a range of imaging components most of which are situated in a 'rotating ball' on its nose (Singer, 2011, p. 33). Within this and the body of the Reaper are 'an infrared (IR) sensor, a colour/monochrome daylight electro-optical (EO) TV and an image-intensified TV' as well as 'Synthetic Aperture Radar (SAR) and Ground Moving Target Indicator (GMTI) to provide an all-weather capability' (Royal Air Force, 2016). These capabilities build on earlier efforts that have either enabled wide angle information collection in low resolution or narrow fields of view in high resolution.[16] As noted above, the development of new technologies for Reaper has enhanced the vision so that, depending on the amount of zoom you can either make out the type of rifle someone is carrying or you can survey an area 'much wider than Chipping Norton'.[17] With this capacity to see at different ranges there comes the need to interpret what is being see in different ways, to make or unmake patterns, to infer, to make 'truth' from the 1s and 0s sent via cable, via satellite. Peter Lee describes a scenario:

> The vehicle begins to track slowly through the back streets. 'The driver is looking for something or somewhere but probably does not know the area well.' The Boss interprets what we are seeing for my benefit. Then it stops, reverses and moves backwards and forwards a couple of times as the driver tries to get closer to the adjacent building. He ends up further away than when he started. 'This is an Austin Powers parking manoeuvre right now,' says the pilot.
>
> (2018, p. 41)

Here what is seen, what 'the Boss' sees is interpreted in a particular way – based perhaps on prior watching, on experience, on previous assumptions made: but whatever the background, this becomes a kind of truth in ways that 'erases the particularities of seeing, including how footage from drones is being used, the sorts of people interpreting it as truthful in specific contexts and how given footage achieves its effects' (Grayson and Mawdsley, 2018, p. 12). But this is also an example of the way in which the crews engage not (just) with masculinist logics of seeing but are also required to implement subjective assessments to create knowledges, in order to make sense of the data collected by the drone – something this chapter will go on to unpack. Here the queer logic(s) of Haunting introduce the instability at the centre of relying on masculine rationality alone in warfare. Warfare is a strange and fluid phenomenon, as Clausewitz's oft quoted reference to the 'fog of war' gives hint to. In the way that this fog alludes to a spectral absent-presence, so too does it indicate the recourse of a singular understanding of gender logics of war are limited. Therefore, the following section goes on to explain the backdrop of rationality through which the crews are haunted by the phantoms of 'other' knowledge, of hunches, intuitions, the things on the periphery 'shadows, ephemera, energies, ethereal forces, textures, spirit, sensations' (Puar, 2007, p. xx).

(Un)manned gaze

One of the reasons that in/(hyper)visibility is a core concern of Haunting is because ghosts occupy a 'liminal position between visibility and invisibility …' (Blanco and Peeren, 2013, p. 2). They are seen and not quite seen, they see and do not see. Also in/(hyper)visibility is politically important to the concerns of Haunting because being visible is a core condition of 'liberation and empowerment' from the crushing weight of being politically unseen or ignored (Parkins and Karpinski, 2014, p. 4). As noted in Chapter 2, what is seen is constructed as what matters, what we should be concerned with, what we should take seriously. Therefore, it follows that what remains unseen, invisible, are the things we need not be concerned with. However, challenging this perspective is something that is at the very heart of utilising Haunting as a framework, both in terms of methodological and political commitments. By asking what is not seen, what is invisible, it is possible, methodologically, to draw attention to the absences that gender theorising points to as integral to the feminist project. And in building queer logics into the methodological framework, it is possible to break open some of the dichotomous forms of thinking about gender and gendered realities in warfare in novel and useful ways.

Visibility is constructed in feminist thinking in this way in part because of the connection between masculinity and ocularcentric logics, a legacy of the scientific Enlightenment.[18] As such, Parkin and Karpinski argue that the

> coupling of power and invisibility can be traced back to the epistemological conceptions of scientific truth and objectivity, which ignore the knower's situation and construct a universal view, simultaneously from everywhere and from no-where.
>
> (2014, p. 5)

And it is precisely this perspective which, as outlined in Chapter 2, Haunting aims to challenge. By undertaking a ghost hunt the following section indicates the extent to which the crews use of Reaper drones for surveillance is intimately interwoven with ideas of masculinity. These ideas of masculinity are haunted by narratives of domination, omnipotence and sexuality that cannot be captured without paying attention to the sensuous knowledge of ghosts. Employing this knowledge and queer logic(s), and specifically engaging with the concept of the warrior itself, I demonstrate that whilst the arguments about the masculinisation of the crews engaged in persistent surveillance can be seen to be 'true' so too can attendant arguments about feminisation. The queer disorientation of drone warfare, which Daggett (2015) describes as not yet solidified, provides the ethereal backdrop against which claims about the importance of in/(hyper)visibility to thinking about the juncture between gender (politics) and military technologies like drones.

Ocularcentric

Persistent surveillance, as previously noted, is based on the idea of constant watching, of continuously collecting visual data. In their discussion of drones, Shaw and Akhter use similar language to Parkins and Karpinski (2014) to draw attention to the link between surveillance, contemporary militaries, and masculinity noting that

> Vision is ... crucial to an ocularcentric Western society ... and always already entangled within military culture. The ability to gaze from 'nowhere' and yet represent 'everywhere' ... is fundamentally located within a nexus of disembodiment.
>
> (2012, p. 1495)

These authors write for very different literatures, and perhaps audiences – the crossover in the language used illuminates the threads of gendered thinking which weaves through discourses on military operations. Indeed, constructing the Western military vision as disembodied and as situated 'nowhere' reflects the masculinist logics of the Enlightenment period in eighteenth-century Europe.[19] During this period there was a move towards a preference for visual verification of 'facts' as part of a desire to 'ground knowledge in verifiable, empirical data' (Douthwaite, 2002, p. 72).[20] There is not space here to outline the Enlightenment story, which has been extensively explored elsewhere, but suffice to say that, as part of the heritage of this important period of intellectual development, social science (as well as the natural sciences) came to embrace the idea that the simplest means of ensuring that data was 'verifiable' was to rely on what could be seen, so that the 'truth' became what the eyes verified as 'reality'.[21] In addition to constructing what was seen as what was real, seeing became connected with ideas of understanding. What we see gives us more than just visual cues, through seeing we claim we 'understand' (Lyon, 2014, p. 23). Or, to paraphrase the tagline of the 1996 film *Loch Ness*, when we see we believe. And therefore what cannot be seen cannot be thought of as 'true' or 'real' and the unseen is then rendered 'unreal' (or unproved). Not only was seeing clearly connected to arguing for what is real, is empirical fact, is, perhaps, truth. But also, as per the connection between Shaw and Akhter's (2012) description and Parkins and Karpinski's (2014), this fact, this understanding of truth is connected which a rationality clearly constructed as masculine. Widely expounded during the Enlightenment period, this construction of masculine knowledge as rational and feminine knowledge as irrational (transposed so often without thinking onto the bodies of those identified as male and female or man and woman) continues to affect constructions of what constitutes knowledge.

If arguing that 'seeing' provides an 'objective' means through which to access knowledge is one component of the connection between vision and masculinity, then the second is the hierarchical power relationship between the individual who is 'seeing' and the individual who is 'seen'. Importantly, for the connection between seeing/understanding and masculinity 'Historically, man has been the subject, the agent "doing the looking", whilst woman has been the object of his gaze, the spectacle' (Rowley, 2010, p. 312).[22] The relationship between the 'seer' and the 'seen' and gendered hierarchies is also (hyper)visible through the active nature of seeing (activity being coded as masculine) and the passive nature of being seen (passivity being coded as feminine). Reflecting masculine agency, the male gaze is not without effect, but rather has political power and implications: 'Masculine vision is almost invariably characterized as patriarchal, ideological, and phallocentric' (Snow, 1989, p. 30). To see in this way, through a masculinist lens, the argument goes, is therefore far from objective but overlaid with judgement that derides and dismisses the feminine with the aim of dominating. Therefore, to be the recipient of/object of the gaze is to be feminised, and as such denied agency and placed in a weak position vis-à-vis the watcher.[23]

In drone warfare, the position of the Reaper crews as 'above' the individuals that they watch acts as a physical representation of the hierarchical relationship between the watcher and the watched. As Allison Williams notes, 'the aerial view as a given, neutral and all-seeing, is contested ... and the problematic positioning of the geopolitical gaze as a subjective, located, practice can be uncovered' so that awareness can be bought to the raced, gendered, sexualised power dynamics at play in drone warfare (2011, p. 384).[24] The 'politics of verticality', where the watcher is placed physically above the watched perhaps renders this hierarchy at its most obvious.[25] The perspective of seeing from above by drone crews creates a specific way of seeing the world (Stahl, 2013; Gregory, 2011, p. 190; Kindervater, 2016). Some academics (Stahl, 2013, p. 663) argue that 'drone vision represents a special kind of looking' in that the capacity for persistent surveillance, ability to see widely and in detail, conjures what Donna Haraway describes as the 'god trick' (Haraway, 1988). Haraway's 'god trick' questions the idea of 'seeing everything from nowhere', arguing that all knowledge is situated and that, therefore, strict objectivity is not achievable, either in (social) scientific study nor, for example, in the conduct of warfare. No matter how much data is collated, no matter that 'vision in this technological feast becomes unregulated gluttony', it remains impossible for the eye to see all in the sense of being able to know all, this is only a masculinist myth (Haraway, 1988, p. 581). What I argue in the coming sections is that in the visual gluttony of drone warfare, because there is an unsettling sense that it is not (never) possible to see and know everything, that some reliance must (secretly, and unseen) be placed on other knowledges, non-visual and feminine knowledges. But first, it is necessary to unpack the way in which concepts of visual objectivity are laced with ideas of omnipresence, and through further reference to the deity in the 'god trick' the web of power associated with omniscience.

Omniscience is usually considered an attribute of deities: the capacity to 'know all' is understood as something beyond human capabilities, at least in part because it is not possible to *see* all. As a result the symbol of the eye is widely associated with God(s).[26] Beyond seeing actions, gods' eyes are understood as being able to see into a person's thoughts and emotions, as Da Vinci noted 'the eyes are the window of the soul' (cited in Harper Hart, 1949, p. 1). God(s), it seems, understands the *hidden* desires/actions of the individuals below creating a hierarchical relationship, not least because whilst God *sees* all (what the sinner does is *visible*), S/He cannot be seen (the powerful God is *invisible*): a perspective apparently mimicked by Reaper crews.[27] The connection between God's eye and the technical capabilities of the Reaper are often referred to in narratives about drones, with commentators noting 'they have assumed the position of the gods who decide who will live and die' (Evangelista *et al.*, 2014, p. 193).[28] Additionally, drone-related technology called 'Gorgon Stare' and ARGUS render explicit the desire of marketers of this technology to imbue Reaper with the capacity (of a god or monster with more-than-human powers) to see all: Gorgons being the monsters whose multi-eyed stare could turn enemies to stone (controlling the behaviour of those it could see, much as Grayson and Mawdsley (2018) argue drones do), and Argus (specifically Panoptes) is the many eyed giant of ancient Greek mythology who was described as 'all-seeing' and therefore with monstrous powers. The connections between gods (and/or monsters with god-like powers) extends beyond Western interpretations of drone warfare. Indeed, to the people who live in the areas of Waziristan, where US Reapers operate, drones are compared with 'Ababels (The holy swallows sent by God to avenge Abraham, the intended conqueror of the Khana Kaaba)'.[29] This conceptualising of drones, by potential victims of drone strikes, as connected with the gods and with omniscience (perhaps even omnipotence) is a powerful illustration of one element of the masculinisation of the Reaper crews.

As illustrated in the previous section objectivity is illusory because watching and being watched are not passive, nor equal, positions. The eye that 'sees all' is empowered over (and above) the one that is watched. As the application of Haunting reveals, the process of surveillance, and what is seen through surveillance, is a form of 'sensuous' 'situated knowledge'.[30] As noted before, 'situated knowledge' refers to the feminist understanding that knowledge is not objective because it reflects a specific standpoint, the 'situation' or 'situatedness' of the individual who produces it, based on/in their cultural, social, historical, psychological and personal biases and perspectives on the topic.[31] In drone warfare this plays out in how data is collected, interpreted, made into 'knowledge' of those being watched – in how their activities are categorised as threatening or benign – as if that dichotomy could somehow capture the full panoply of human behaviours and their interpretation from a distance.

It would appear that, so far, this appears to reflect a traditional and consistent representation of military masculinity. What then contests this representation? What are these experiences haunted by that are best explained as feminine or through queer logic(s)? It is, perhaps, attention to the invisible, or perhaps the failure to make knowledge from what is visible. For example, the limits of surveillance to provide god-like 'knowledge' of what is seen below is revealed in the following excerpt from US Reaper operator Martin,

> Our supported unit in Afghanistan wanted me to keep an eye on [a mound] until daybreak … I would continue to circle and stare at it…. Hours passed…. Once the sun rose … I asked my sensor operator to switch to daylight cameras. The telephoto lenses brought out the full picture … I burst into laughter. The feature we had stared at for hours, that had preoccupied the most sophisticated reconnaissance apparatus on the earth and baffled the world's finest intelligence analysts, was a pile of barnyard manure.
>
> (2010, pp. 33–34)

Here is an instance where there is a failure of 'particular artefacts [to be] … invested with truth-telling capabilities' (Grayson and Mawdsley, 2018, p. 5). There is a failure of the visual, the penetrative capabilities of the drone's cameras to unveil the object. Whilst this failure is evidence of the limitations of drone technologies, it also points to the problems with these limitations. After all, this pile of steaming manure is not watched for hours by a crew for the sake of watching. Rather it is watched because in its partial invisibility, its indecipherability, it is constructed as dangerous and this can have, in other scenarios, potentially life-threatening implications. In this instance the manure that could not be interpreted in the dark, *could* be interpreted in the light, could be seen and made sense of. But what if that had not been the case?

Whilst one interviewee claimed that there was 'something incredibly seductive about trying to develop your understanding of what is going on ... it's not a story, it's what's going on', this is perhaps not entirely accurate.[32] What is going on *is* going on, but whether or not that is seen and understood by the crews watching depends, at least in part, on their ability to make a narrative, a story, about what they are seeing. This steaming pile of manure would exist whether or not it was watched by Reaper crews. The interpretation of that pile is what matters for military decision-making, and I would argue, that interpretation is necessarily that: interpretation. It therefore necessarily falls short of the masculinist rational of constructing 'truth' from the visual, suggesting that other logics need to be explored here.

If you read the transcripts from the US civilian casualty incident published by David Cloud about whether what is seen are women or MAMs (military aged males) in 'man dresses' both indicate the limits of interpreting surveillance data and the gendered nature of that interpretation – with women assumed innocent and MAMs assumed or interpreted as guilty. This is explored by Lauren Wilcox who argues 'The practice of "seeing" only "military-aged men" through the digitized processes of the drone assemblage works on the basis of a resemblance' (Wilcox, 2017, p. 12). The queerness of the enemy expressed through confusion over men in dresses is further expressed in the following segment from Cloud's (2011) transcript:

SAFETY OBSERVER: Are they wearing burqas?

SENSOR: That's what it looks like.

PILOT: They were all PIDed [positively identified] as males. No females in the group.

SENSOR: That guy looks like he's wearing jewelry and stuff like a girl, but he ain't ... if he's a girl, he's a big one.

In a similar UK scenario, Reaper crews struck two vehicles laden with (perhaps) homemade IEDs, in what was described as a 'Textbook scenario. Isolated, middle of nowhere. No collateral risk' (Lee, 2018, p. 99). But after the strike is made (as described by Lee): 'A slight movement through the haze in the picture caught Jamie's attention.... Jamie's stomach lurches. Next to the man is a person half his size. "Is this ... is this ... a CHILD?"' it is later confirmed that amongst the dead is at least one adult woman and one child, at that point 'identified easily' (Lee, 2018, pp. 101, 105).

Examples like this invite a question of the omniscience of the crews. The masculine capability to *know* or to extract knowledge (without permission from the watched individual) from what is seen here dissipates, revealing a feminine need to rely on hunches, suspicions and emotional knowledge. This plays out in life or death scenarios, for example, crews have needed to interpret a 'flutter' in the corner of the screen as a group of school children in order to prevent a strike that would have killed the entire group.[33] Here what was *seen* was not a group of children, rather it was a ghostly movement in the peripheries of vision, and as such crews not only needed to take this spectral form seriously, but also to engage with intuitive forms of knowledge that have traditionally been constructed as feminine. Even apparently masculine 'data' collected is described as 'patterns', as in 'patterns of life' intelligence that it used to determine the threat posed (read: guilt) of the individual(s) being watched. This determination based on patterns and probabilities, based on suppositions and assumptions about normal and abnormal behaviour, also relies on those kinds of knowledge(s) described or interpreted as feminine.

For example, one of the forms that sensuous knowledge takes is through the designation of something as 'uncanny', the familiar rendered strange.[34] This plays out in the lives of Reaper crews through an acknowledgement of the connection between the lives of those designated as 'targets' and those tasked watching and targeting them. In watching the domestic lives of the individuals below them, in seeing the similarities and differences, Reaper crews experience moments 'when familiar words and things transmute into the most sinister of weapons and meanings', which impacts not just on their perspectives of what is occurring on the ground but also on how they interact with their own families (as will be addressed in greater detail in the next chapter) (Gordon, 2008, p. 64). Examples of viewing the domestic as strange or dangerous include confusing digging irrigation channels for the planting of IEDs, the pile of barnyard manure viewed as a mysterious/dangerous 'heat spot', and 'flutters' in peripheral vision coalescing into a group of children. These things are simultaneously strange and familiar, the lives of the targets are like the crews' own and at the same time completely different, the people who are watched are simultaneously other and 'never that' (Gordon, 2008, p. 5). Therefore, whatever the 'power' of the Reaper to stay above the targets and watch and watch it is impossible for the crews to ever be able to fully understand, in the way that a god might, what is the 'real' story. Their assessments are always partial, fluid, and subjective.

In addition to highlighting the limitations of crews' abilities to turn what is visible and what is seen into knowledge, the experience of persistent surveillance can force Reaper crews to engage with 'sensuous knowledges' through the powerful emotions that their experience when they see upsetting, troubling or anxiety-inducing visions. Whilst the crews are operating at a vast physical, geographical distance from the theatre of war, they are not as distanced from the emotional and physical responses to experiencing war as some commentators' claim. As MP David Davis described

> They witness the violence, whereas most people who are involved in a war are distanced, at least to some extent, from the people who suffer from their weapon system. The sheer fidelity of the drone systems makes the witnessing very close and personal.
>
> (Davis in Hansard, 2015 Column 50WH)

It is this trauma and concurrent intimacy of surveillance via Reaper drones and the implications for the work that gender does in this context, to which this chapter now turns.

Intimate in/visibility

One of the roots of Haunting as a framework comes from an investigation of trauma and memory and there is increasing interest in the spectralities literature in how this trauma emerges through discussions of in/(hyper)visibility.[35] As part of the experience of persistent surveillance Reaper crews have disturbing images rendered visible in three ways: (1) in the operational requirement for battle damage assessments (BDA) where crews have to assess the aftermath of weapons strikes, counting bodies (or body parts) in order to decide whether or not operational targets were met; (2) in the observation of the death of friendly forces or civilians; (3) in the observation of traumatic 'social events' that the crews are not permitted to intervene in. As a result, these crews may be haunted by the experience of becoming distantly intimate with the subjects of surveillance.[36] Contrary to critics' claims that the physical distance between the crews and the theatre of war will lead to a moral disconnect, crews are instead indicating that they are troubled by unexpected emotions, disturbed by the ghostly similarity between their lives and those of the individuals under surveillance.[37] 'Jamie' who spoke about his experience of causing civilian casualties noted watching an older man coming up the road to where the bodies of the dead were covered with tarpaulin 'He lifted the tarpaulin and looked for a good few seconds. What was he thinking? Was he praying? He bent down and lifted the body of one of the children – there were two – and walked slowly away'. Jamie then recounted that after picking his family up from a barbeque that evening, one of the children falls asleep in the car: 'It wouldn't be the first time he had had to carry a sleeping child from the car into their bed. [But] Images of an elderly Afghan carrying a child in the dark' haunt him (Lee, 2018, pp. 106, 108).

In another example, worth quoting at length: After destroying one enemy vehicle

> … these two guys came back to inspect what happened. I felt sorry for them. I felt some association with them because my dad used to take me for drives in trucks where he used to work. I felt an affinity with them. Then we were 'cleared hot' on them. We rifled them [used a Hellfire]. They heard the sound of the missile as it approached, then one of them threw himself on top of the other. And that's just stuck in my head. I don't know why it's affected me. I don't know if it was the affinity with my father and the truck. I believe they were father and son because of the way the one threw himself on top of the other. And they were both lying there, dead. It's just something that's always stuck with me. They were just men like me.

> (Lee, 2018, p. 224)

Here the Reaper crew member makes an assessment, the kind of construction of truth discussed earlier as a means of disciplining the data into (rationalist, masculine) knowledge, which bridges the divide between the rational and the emotional, the factual and the fictional, the (remembered) past with his own father and the present (lack of future) of the deceased possibly-father-and-son on the ground. The Hauntological framework here destabilises both the narratives of cold, impassive, stoic professional of contemporary military narratives and hot-headed emotional and passionate warrior of myth, destabilising the masculine and feminine constructions that act as anchors to both of those constructs. What is present in this process of seeing, what is not see-able, what is invisible (i.e. the true nature of the relationship between the two men), is not merely both masculine and feminine and therefore not gendered, but rather something which destabilises the distinctive categories to provide something which is *simultaneously* masculine and feminine, deeply gendered.

These claims destabilise traditional narratives about geographical distance and emotional connection, and in so doing run counter to arguments that drones act as a masculinising influence, reducing the warriors' access to the emotions that make war a human endeavour.[38] The experience of seeing violent, horrifying, and traumatic things has implications for our understanding of the masculine as emotionally stoic and the position of rationality as a trait privileged by the warrior. By drawing queer logic into the mix, the ghost hunt of the trauma of surveillance and the curious sense of intimacy that *persistent* surveillance can create reveals a more nuanced picture than we might initially expect. Indeed, at the end of his extended argument against the use of armed drones (as they are currently used by the United States military and, particularly, intelligence agencies), Pryer includes a section titled 'On the Importance of Appearing Human'. Whilst he is using this section to outline the way drone warfare dehumanises the enemy and the country using the technology, it appears that he has missed the way in which his argument actually humanises the drone and loses sight of the real flesh-and-blood humans who are involved (Pryer, 2013, p. 21).

The visibility of the 'Other'

In most assessments of the experiences of Reaper crews they are positioned as the masculine subject 'watcher', rather than the feminised object of the 'watched'.[39] What many of these accounts touch on, but fail to really engage with, is the way that the crews' subjective assessment of the individuals that their surveil results in a strange intimacy. This intimacy is ghostly because it is simultaneously a real intimacy and also an unreal intimacy. The crews know the pattern of life of their targets, they build up pictures of complex webs of relationships within communities, and they get to know who is a 'stranger' from the things that they *see*, the things that are visible. However, this intimacy is also *un*real because the relationship is one-directional, the targets do not necessarily know that they are being watched, and the knowledge that the crews can make *visible* still cannot truly establish intentions, motivations, meanings behind certain behaviours. Therefore, however god-like the position of power above the ground might seem, the Reaper crews, unlike God(s) do not have a window into the soul. That impression is merely a god-trick.

As noted earlier in this chapter, one of the ways that this strange intimacy haunts the Reaper crews is through the apparent similarities between the lives of their targets and their own lives. Reaper crews observe 'women hanging out the washing or children playing', '[owners] playing with their dogs', 'a group of kids throwing rocks at goats, [a target] mak[ing] love to his wife in an open field, attend[ing] a wedding or funeral' (Freeman, 2015; Abe, 2012; Hurwitz, 2013; Daggett, 2015, p. 371). All of these minutiae of everyday life serves to disrupt the traditional 'Othering' of the enemy in war: his/her life appears not so different from the crews', his/her family has similarities to the crews'. For example, compare the day-to-day human experiences of the 'targets' listed above with those of the crews: interviewees spoke about 'rushing home after stressful shifts to do the housework',[40] another noted he'd be checking the weather in Kandahar whilst watching TV with the kids,[41] one who described the camaraderie: 'we swear a bit, drink a bit and are a bit rough ...' pointing to subjectivity made out of more than *just* being a Reaper crew member.

Furthermore, the knowledge of the (not so) 'Other' accumulates in a strange way so that one pilot noted 'Sometimes I thought I knew as much about this man, my nemesis, as I knew about my wife' (Martin, 2010, p. 296). Again, Martin connects the idea of watching/seeing with the idea of knowing, and interestingly with an intimate knowledge. To *know* someone in the biblical sense is to know them sexually, in the manner often considered to be most intimate. In sexual relationships we expose parts of ourselves (both physically and emotionally) that we do not share with other people. Martin's claim to know his enemy as well as his wife then serves a useful means of illustrating the way in which Reaper crews' connection with the individuals that they watch serves to trouble traditional ways of thinking about distance in war: 'And they were both lying there, dead. It's just something that's always stuck with me. They were just men like me' (cited in Lee, 2018, p. 224).

Grossman (2009) argues that the greater the physical distance from the target the lower the resistance to killing. If this hypothesis held then drone crews would have an easy job hitting their targets, emotionally (as much as physically) distanced from the consequences of their actions. Indeed this is frequent criticism of the use of drones.[42] However, Chamayou problematises this perspective asking, 'Where should the drone be positioned in the diagram?' (Chamayou and Lloyd, 2015, p. 115) to which Daggett responds

> Arguments could be made in both directions: there is a maximal distance between shooter and target, putting drones beyond long-range bombers, but at the same time there is an odd intimacy made possible by the drone cameras and surveillance capabilities.
>
> (2015, p. 366)

In addition, if we consider Martin's claims and the connection between his 'knowledge' of his enemy and sexuality then far from being at the extreme distance from his target, the Reaper pilot/sensor is placed at the point at which, according to Grossman there is likely to be the *greatest* resistance to killing. It is not simple to separate these experiences: the distance and intimacy are interwoven in narratives of watching, killing, closeness, emotion and identity. As interviewee Dan noted

> It's a long, prolonged exposure to the guy: you will find him ... [there is] much more emotional involvement with the target ... [it] affects you more than it does turning up at a target [and just] hitting a target ... [Even though] the physical visuals are the same.[43]

Dan's statement draws attention to the usefulness of Haunting for investigating the intersection between gender and drone warfare – particularly, here, the thematic concern of in/(hyper)visibility and its relationship with complex personhood.

This first half of the chapter has sketched out the gendered implications of the role of Reaper crews as watchers. I have illustrated how this position can be construed as both masculinising and empowering, *and at the same time* feminising and traumatising. The following section then reflects the second component of persistent surveillance in the world of the Reaper crew, and that is the experience of being watched.

The watchman and the politics of (hyper)visibility

Reaper crews are usually invisible to their targets, but they are distinctly visible to military leadership through a web of connectivity as this section will explore. However, the crews have also been *invisible* in early official reports and newspaper articles about drones and particularly drone strikes,[44] until they became constructed as this new frightening breed of warrior and rendered *hyper*visible.[45] Utilising feminist thinking on the way that visibility in these different forms and weaving this together with Gordon's concerns about how visibility dis/empowers, this section traces the webs of power at work in the crews' experiences of being scrutinised.[46] Through this I argue that the crews are *simultaneously* being feminised by their construction as (invisibly) passive and in need of supervision *and* provided with the opportunity to perform military masculinity in a manner that reflects the queer logic(s) at play.

The spectre of the cyborg

One of the things that struck me when I became interested in the phenomenon of military drones was the way the narratives reflected a sense of the drone as the actor, despite the fact that one of the core claims of the using militaries was that these machines are under human control.[47] Particularly at the beginning of the debate over the ethical, legal, and strategic implications of using armed drones there was little reference made to the crews behind the aircraft. Newspapers reported that 'British drones kill hundreds of Taliban fighters', 'drones kill innocent people all the time' and that drones are 'a weapon capable of finding and killing someone just about anywhere in the world' as though the technology itself was a cognisant actor (Rayment, 2014; Bowden, 2013; Beckhusen and Gault, 2015). I appreciate that some of this invisibility of the crews may be a result of crafting a good newspaper headline, however the sense that it is the *drones* that are capable and active pervades even commentary from the using militaries. For example, the former Assistant Chief of Air Staff noted that 'they have unique capabilities' and what he is talking about, what constitutes the 'they' is the drone machinery itself (Whetham, 2013, p. 22). Whilst there is a long history within the British military of naming aircraft and pilots/crews expressing a sense of connection with their aircraft, with drones this plays into discourses of autonomy, about cyborgs and of the drone enacting warfare without human intervention.[48] In these iterations it is the drone itself that represents the cyborg and poses a threat to normal human relations through the autonomous deployment of lethal force, creating images of dystopian, *The Terminator* and *The Matrix* inspired futures where 'robotic killing machines … systematically hunt down human beings' (Hasian, 2016, p. 106; see also Turse and Engelhardt, 2012). This perspective has also managed to gain some currency within the militaries using Reaper drones themselves. For example, in the 2013 issue of *Military Review* Lt Col Douglas Pryer (US Army) included an image of 'armed terminator robots and hovering drones fighting humans in a scene from the movie *Terminator 3: Rise of the Machines?*' in a piece that critiqued the use of unmanned/robotic platforms as a danger to the United States (Hasian, 2016, p. 5; Pryer, 2013). In the piece Pryer describes Pakistan and Yemen as 'the locales in which armed Predators and Reapers hunt', describing 'the ever-watchful eyes of armed drones', which again, give agency to the machines themselves, making them animalistic rather than a technological object and erasing the crews that are tasked with operating them (Pryer, 2013, p. 17).

Reaper crews have expressed frustration with this erasure of the work that they do, which results in, one interviewee claimed, 'a grand, grand underappreciation' of the work that they did and do.[49] Speaking about the air strikes in Syria, Dan expressed irritation that 'all the press in Syria is a good example, always a picture of a tornado … we've been there for ages and we've done so much more than them …'.[50] This invisibility renders their work apparently unimportant, lesser or not worth notice, because, as discussed at the start of this chapter, what is seen is what is considered important.

Following a period of invisibility, the crews then became the subject of much more interest. As the press in particular bored with just addressing the (still contentious and incredibly important) issues of legality and ethics, the shadowy figures behind the technology began to materialise in reports. However, rather than doing an important job of rendering this previously invisible individuals *visible*, I argue they were rendered *hyper*visible (as described in Chapter 2). The crews were rendered *hyper*visible in later accounts because they were con-structed as strange, as different from the crews of manned aircraft, as 'Other' and therefore possibly dangerous. Headlines for these pieces included 'Confes-sions of a drones pilot', 'Armchair killers' 'Killing by drone and proud of it', 'The dark art of drone piloting', 'A Candid, Chilling, Conversation with Top US Drone Pilot' (Power, 2013; Freeman, 2015; Rattansi, 2010; Reynolds, 2015; Wood, 2013). The uses of terms like 'confessions' implies that some grievous crime or sin has taken place that the individual needs to atone for. References to 'dark arts' (almost literally) conjure up imagery of black magic, demons and devils that Reaper crews are supposedly using to operate their aircraft (rather than good old-fashioned professional training). Unlike the crews of fighter jets or helicopters, the Reaper crews are flying something inherently devilish or demonic (admittedly hindered by the names 'Predator' and 'Reaper'!), which means that journalists engaging with these crews are involved in something 'chilling'. This kind of language clearly suggests that the crews are involved in something morally and ethically reprehensible (as many of these articles do). But it is not only the ethics of the pilots' roles that are held up for questioning.

In an article dripping with disdain Calhoun claims that drone pilots can 'run out for a Starbucks break in between their various point-and-click killing mis-sions' and explores how the use of drones signals the 'end of virtue' in military endeavours (as explored in Chapter 4) (2011, p. 379).[51] Drawing on the heritage of the ghostly warrior (of Chapter 3) commentators have critiqued, satirised and poked fun at these 'armchair killers', constructing them as the very antithesis of the brave, manly infantry or special ops officer: something that has not gone un-noted by the crews themselves who bemoan being made to 'look like twats on TV'.[52] Consider the following cartoon:

Figure 5.1
Source: Sinann, 2012.

The 'traditional' Top Gun fighter pilot is taller and slimmer than the drone pilot, he smiles in a way reminiscent of movie stars, and wears 'cool' sunglasses. In comparison the drone pilot is too fat to fit into his t-shirt, he is wearing trainers, and glasses (implying that his eye sight is less than the perfect required for pilots), and he sips a soda (another reminder of his less than peak physical fitness). The small stature of the drone pilot here serves not only to emasculate him (men are generally considered to be taller and physically 'bigger' than women (Goldstein, 2001; Van Creveld, 2001)) but also to infantilise him. Being smaller than the fighter pilot he appears childlike, a perception reinforced by his casual clothing.

Situating the drone pilot as childish serves to feminise him. Feminist scholars have noted how women are often placed together with children in security issues to create an amorphous mass that Enloe (1989) refers to as 'womenandchildren'. Whilst a number of gender scholars have problematised this connection between women and children, the idea that 'that children are not fully mature, are depicted as not capable of rational thought and are also seen to be in need of care and protection' (Shepherd, 2009, p. 41) is also highly problematic (and I would argue, inaccurate) when applied to individuals who are responsible for multi-million pound aircraft with the capacity to undertake lethal strikes.

The world's longest screwdriver: the ghost of the commander

This final section of this chapter addresses the meaning behind the chapter title, that is, the claim that crewing Reaper is like being part of 'the longest screwdriver in the world'.[53] Here the spectre of public invisibility or *hyper*visibility is displaced by the connectivity of the Reaper aircraft, as crews comment on concerns that phantom military leadership will climb into the cockpit with them to oversee (and engage in/direct) their operations. Occupying the feminised position of being watched, the scrutiny under which Reaper crews operate creates a space of queer logic where they are also provided with the audience to which to perform military masculinity understood through their mastery of the complex technology of the aircraft.

Oversight is part and parcel of the lives of the majority of members of the British military, but one of the things that makes the Reaper novel is that the feeds that they capture can be reviewed not just by the crews but by other members of the military, and occasionally by other agencies (for example intelligence) and by civilian leadership (prime ministers/presidents, etc.). All of this adds up to a very different crewing experience from that of other aircraft in which the individual or crew, once tasked, is responsible for how they undertake that task and they have the freedom to make decisions, largely unseen and unwatched by superiors. Built into this freedom is a professional trust that means that, once trained, military leadership are confident that the individuals that they task will do the job professionally and within the bounds of the laws of war.[54] One interviewee recounted that in his previous role as an Apache pilot it was 'just me and the [co]pilot' and that they had to 'get on with it, just the two of us'.[55] In this scenario the individual's identity was partially built around his independence, his capacity to operate his aircraft in a complex environment, and his ability to make difficult decisions in potentially life-threatening scenarios; all of this safe in the knowledge that his action would (provided he didn't act recklessly or illegally) be judged good by his peers and superiors. He had the freedom (and the responsibility) to 'get on with it', a position that reflects the independence of mind (as a component of masculinity) demonstrated by the figure of the warrior through iterations such as Achilles and Odysseus.

Reaper crews are different. Described as the 'most supervised strike platform bar none' Reaper crews operate in an environment of extreme connectivity.[56] Forty people may be connected to one crew, able to see, watch, speak to, ask questions of, make demands of, contradict, perhaps even confuse: that is a world away from a position of being 'just the two of us' who were empowered to 'get on with it'; and represents a cultural step change from previous iterations of the pilot who was alone in the cockpit with 'no man stand[ing] at their shoulders to support them' (Gen Billy Mitchell cited in Cantwell, 2009, p. 75).

The implications of this close scrutiny are interesting. On a simplistic level, this overwatch puts additional pressure on already stressed crews. Noel was clear that being over-supervised can have negative effects noting that 'there's a real risk with this long screwdriver … it's really appealing [but] … we mustn't allow our senior commanders to get back in the cockpit' because this negates the professional abilities and identities of the crew members.[57] Because there is no individual inside the cockpit of the Reaper, crews are already fighting to be acknowledged as an important part of the Royal Air Force, and, as previously noted, there is a tendency within both journalistic and official discourses to anthropomorphise the drones, denying the involvement of any military staff in their use. Maintaining a 'long screwdriver' that puts apparently phantom commanders 'back in the cockpit' actually serves to remove the crews themselves. Being closely supervised, then, is not only frustrating on a day-to-day level but also leads to a perception that the crews are less masculine, less capable, less professionally trustworthy than their colleagues.

Whilst all of this serves to suggest that the supervision and scrutiny of the Reaper crews can be perceived as being wholly feminising, I want to use the following section to describe the ways in which, whilst this is true, it is *also* true that it serves to masculinise the same crews. The haunting of the crews by the eyes of their superiors and colleagues provides the platform or stage upon which they can demonstrate their technical mastery of the complex technology that they are tasked with operating, and as previously noted, technological mastery is construed as a masculine trait.[58] As noted in Chapter 1, I understand gender in this book not as something you have but as something that you do. This is not to argue that choosing or changing gender is simple, it is most certainly not, but rather that gender is something performative that must be constantly maintained through re-iterations of acts (broadly understood).[59] Therefore, doing the work constructed as masculine and being seen to do that work is an important part of being identified by self and others as masculine.

To this end, interviewees noted that the Reaper is 'not an easy' aircraft to pilot, and therefore being able to do so without 'pranging' it was an important marker of masculinity.[60] Understanding technological mastery as a masculine trait, Freddie described conducting accurate strikes as 'surgery from 10,000 ft', implying control, finesse, and the skill of the crews. Similarly, interviewees noted that one of the reasons they had applied for the role or enjoyed the job was because 'UAVs are the way forward' so that operating them represented 'being on that leading edge, the crest of the wave' of military technological development and learning.[61] Because of the operational tempo, in addition to learning to use 'cutting edge' military technologies, the crews were also permitted (and required) to push the aircraft to its very limits. This is different to other crews where they might practise the most complex moves again and again and never have the opportunity to use them operationally. As Tom recalled, when he was working with Tornados and Typhoons 'we practised rarely used complex bits but in Reaper we *did* the complex bits!' and importantly, they were *seen* doing these 'complex bits'.[62]

Again, whilst the mastery of technology is an important marker of masculinity the recognition of this requires an audience, and that is where being so closely scrutinised provides a means of masculinising the crews. Much like the oppositional relationship between masculinity and femininity, the different kinds of masculinities interact relationally and hierarchically.[63] Therefore, in order to claim a higher order of masculinity, the crews needed to demonstrate their status as *higher than* another's. As Robert recalled, the 'rest of the Air Force, particularly the fast jet boys, looked down their noses at us' but 'we were providing better situational awareness' than they could, an operational skill that provided a means of performing military masculinity.[64] The importance of these skills being seen to how the crews were viewed by others is illustrated by Dan who noted that when his colleagues in other parts of the forces where able to see the 'amount of stuff I could do from my platform' it 'blew their minds'.[65]

Conclusion

This chapter has utilised Haunting's concern with in/(hyper)visibility to investigate the gendered impact of Reaper crews' roles in persistent surveillance. Beginning by situating the crews in the privileged, and powerfully masculine position of watchers, this chapter has introduced the way this hierarchical positioning is based on Enlightenment thinking which privileges the visual as a source of information. Critiquing this thinking I introduced the challenges that emerge through a ghost hunt of the Reaper crews' experiences of surveillance operations. The 'politics of verticality' on which Reaper crews' apparent position of strength operates reveals itself not to be an objective assessment of data but as relying on 'sensuous knowledges'.[66] These knowledges play out in two ways: first, the subjective judgement of the intentions of the targets on the ground whose behaviour is rendered visible but whose intentions and motivations remain hidden; and second, through the need to interpret amorphous visual data to make *sense* of the shapes, shadows and hunches.

Moving from positioning the crews as the watchers this chapter then considered their experiences of being watched, tracing the ways the press (in particular) engaged with the crews. This initially resulted in rendering the crews invisible, written out of the debate as the drone itself was anthropomorphised and given agency. However, once the debates on the ethics and legality of drone strikes began to pall there was a move that rendered the crews as *hyper*-visible: so that they were registered as strange and monstrous 'Others', a world apart from the professional honourable crews of the rest of the RAF. In addition to questioning the ethics of the crews tasked with using Reaper, I argued that the crews are often feminised and infantilised in discourses and illustrations, compared unfavourably to the individuals who operate manned aircraft.

The crews are visible within the existing structures of military organisation and the way in which this functions to *simultaneously* feminise and masculinise. Interview data and other accounts revealed an unsettled relationship between the crews and other members of the military. The scrutiny over the use of drones has filtered into the cockpits (or Ground Control Stations) of the Reaper, so that in some cases there is a sense of the leadership taking over from the crews who are based there – a spectral overseer that erodes professional trust. Simultaneously, however, the connectivity of the Reaper, the fact that many other individuals can *see* the feeds that the crews collect, provides an audience to which they are able to perform a specific iteration of military masculinity: technological mastery.[67]

Ultimately, what this chapter reveals is that through the prism of Haunting it is possible to see the ways in which Reaper crews' lives are woven through with questions of in/(hyper)visibility. These questions play themselves out through gendered interactions between the crews and the technology of the drone. Being positioned as both watcher and watched situates the crews in the liminal, ghostly spaces between the masculine and feminine, and the use of sensuous knowledges reveals that neither category of watched *or* watcher can fully contain the masculine *or* feminine. Instead, these knowledges demonstrate the way in which the lives of the crews as they engage with persistent surveillance are always already exceeding the gendered binary, demonstrating the importance of including queer logic in addressing this issue.

Both this chapter and the previous one have focused on the way that Haunting can help us to understand Reaper crews lives *at work*, as they operate Reapers. However, as a commitment to complex personhood reminds us, this is not the whole story. The crews are so much more than just their jobs. The following chapter takes this complex personhood seriously by introducing the day-to-day lives of the crews, utilising Haunting to focus on the implications of disturbed temporality.

Notes

1 Interview with 'Robert'.
2 Interview with 'Dan'.
3 Interview with 'Geoff'.
4 Interview with 'Dan'.
5 Noting the same caveat for using the word 'known' as outlined in Chapter 2.
6 (Rowley, 2010; Kaplan, 2000; Conor, 2004).
7 (Williams, 2015).
8 (Freud *et al.*, 2003; Abraham and Torok, 1999; Abraham *et al.*, 2008).
9 Interview with 'Noel'.
10 (Kindervater, 2016; Keegan, 2003; Bousquet, 2008) (Ehrhard, 2010; Wall and Monahan, 2011).
11 (Thomson, 2009).
12 Interview with 'Geoff' and 'David'.
13 Interview with 'Geoff'.
14 Interview with 'Geoff'.
15 Interview with 'Tom'.
16 (Gregory, 2011, p. 193)
17 Interview with 'Geoff', see also comments by David Davis MP (Armed Drones Debate Hansard 2015).
18 (Shaw and Akhter, 2012; Keller, 1986; Haraway, 1988).
19 (Hampson, 1976; Porter, 2001).
20 (Van Loon, 1996; Jordanova, 1999).
21 (for example Hampson, 1976; Porter, 2001; Williams, 1999).
22 (see also Burston and Richardson, 1995; Kaplan, 2000; Brooks, 1993).
23 (Cohn, 1998, p. 144).
24 (see also MacDonald, 2006; MacDonald *et al.*, 2010).
25 (Williams, 2011; Smith, 2016; Wall and Monahan, 2011).
26 (Lyon, 2014; Harper Hart, 1949; Bensimon, 1972; Fingesten, 1959).
27 (Bensimon, 1972, p. 272).
28 (see also Whetham, 2013; Coeckelbergh, 2013; Singer, 2011).
29 (Peshawar Declaration of FATA-based political parties and civil organisations cited in Williams, 2010, p. 884; see also Taj, 2010).
30 (Gordon, 2008; Haraway, 1988, p. 582; see also Weldon, 2006; Kronsell *et al.*, 2006; D'Costa, 2003).
31 (Zalewski, 2006, p. 47; Weldon, 2006, p. 64).
32 Interview with 'Geoff'.
33 Interview with 'Dan'.
34 (Gordon, 2008; Freud *et al.*, 2003).
35 On Trauma and memory: (Freud *et al.*, 2003; Abraham and Torok, 1999; Frosh, 2013); on trauma and visibility: (for example, Engle, 2009; Edkins, 2003; Hawkins, 2010).
36 (Williams, 2015; Abe, 2012; Power, 2013).
37 (Critics include Cole *et al.*, 2010; Alston and Shamsi, 2010; Royakkers and van Est, 2010; Asaro, 2013).
38 (Kunashakaran, 2016; Masters, 2005; Sylvester, 2013).
39 (Williams, 2015; Adey *et al.*, 2014; Asaro, 2013; Wall and Monahan, 2011).
40 Interview with 'Tom'.
41 Interview with 'Robert'.
42 (see for example Cole *et al.*, 2010; Alston and Shamsi, 2010).
43 Interview with 'Dan', similar interview data from British Reaper crews has been recorded by Lee, 2012, p. 17.

44 Invisible in that there were no references made to the crews and the discussions about the use of drones figured the drones themselves as the actors.
45 In this context I use 'hypervisibility' after Shepherd and Sjoberg (2012) to explore how Reaper crews, when acknowledged in reports, are constructed as dangerously 'Other'.
46 (Shepherd and Sjoberg, 2012; Sjoberg, 2012; Parkins and Karpinski, 2014)
47 (Beckhusen and Gault, 2015; Whetham, 2013)
48 (Adams, 2013; Holmes, 2015; Matthew, 2015; *Russia Today*, 2014).
49 Interview with 'Freddie'.
50 Interview with 'Dan'.
51 (see also Baggiarini, 2015; Royakkers and van Est, 2010; Asaro, 2013; Kirkpatrick, 2015; Vallor, 2013).
52 Interview with 'Freddie'.
53 Interview with 'Dan'.
54 Interview with 'Noel'.
55 Interview with 'Dan'.
56 Interview with 'Dan'.
57 Interview with 'Noel'.
58 (Kunashakaran, 2016; Kontour, 2012).
59 (Butler, 1999).
60 Interview with 'Dan' and 'Geoff'.
61 Interview with 'Robert'.
62 Interview with 'Tom'.
63 (Connell, 2005; Barrett, 1996; Higate, 2012).
64 Interview with 'Robert'.
65 Interview with 'Dan'.
66 (Adey *et al.*, 2014; Gordon, 2008).
67 (Kunashakaran, 2016).

References

Abe, N. (2012) Dreams in infrared: the woes of an American drone operator. *Spiegel Online*, 14 December. Available at: www.spiegel.de/international/world/pain-continues-after-war-for-American-drone-pilot-a-872726.html (accessed: 12 November 2014).

Abraham, N., Torok, M. and Rand, N. (2008) *L'écorce et le noyau*. Paris: Flammarion

Abraham, N. and Torok, M. (1999) *Cryptonymie: le verbier de l'homme aux loups*. Champs 425. Paris: Flammarion.

Adams, S. (2013) Calls for ban on development of 'Terminator' robots 'which can select their own targets to kill'. *The Mirror* [online], 13 November. Available at: www.mirror.co.uk/news/technology-science/technology/terminator-robots-calls-ban-drones-2786333 (accessed: 29 June 2015).

Adey, P., Whitehead, M. and Williams, A. (eds) (2014) *From above: war, violence, and verticality* [online]. Oxford: Oxford University Press. Available at: www.oxfordscholarship.com/view/10.1093/acprof:oso/9780199334797.001.0001/acprof-9780199334797 (accessed: 19 February 2016).

Alston, P. and Shamsi, H. (2010) A killer above the law? *Guardian* [online], 8 February. Available at: www.theguardian.com/commentisfree/2010/feb/08/afghanistan-drones-defence-killing (accessed: 4 January 2015).

Asaro, P.M. (2013) The labor of surveillance and bureaucratized killing: new subjectivities of military drone operators. *Social Semiotics*, 23 (2): 196–224. doi: 10.1080/10350330.2013.777591.

Baggiarini, B. (2015) Drone warfare and the limits of sacrifice. *Journal of International Political Theory*, 11 (1): 128–144. doi: 10.1177/1755088214555597.

Barrett, F.J. (1996) The organizational construction of hegemonic masculinity: the case of the US Navy. *Gender, Work & Organization*, 3 (3): 129–142.

Beckhusen, R. and Gault, M. (2015) Drones kill innocent people all the time, but now the White House can't deny it. *War is Boring*. Available at: https://medium.com/war-is-boring/drones-kill-innocent-people-all-the-time-36d5779e92ac#.t8l1iludh (accessed: 22 October 2015).

Bensimon, M. (1972) The significance of eye imagery in the renaissance from Bosch to Montaigne. *Yale French Studies*, 47: 266–290.

Blanco, M. del P. and Peeren, E. (eds) (2013) *The spectralities reader: ghosts and haunting in contemporary cultural theory*. New York: Bloomsbury Academic.

Bousquet, A.J. (2008) *The scientific way of war of warfare: order and chaos on the battlefields of modernity*. London: Hurst & Co. Publ.

Bowden, M. (2013) The killing machines. *The Atlantic*, 14 August. Available at: www.theatlantic.com/magazine/archive/2013/09/the-killing-machines-how-to-think-about-drones/309434/?single_page=true.

Brooks, P. (1993) *Body work: objects of desire in modern narrative*. Cambridge, MA: Harvard University Press.

Burston, P. and Richardson, C. (eds) (1995) *A queer romance: lesbians, gay men, and popular culture*. London; New York: Routledge.

Butler, J. (1999) *Gender trouble feminism and the subversion of identity* [online]. New York: Routledge. Available at: http://kcl.etailer.dpsl.net/home/html/moreinfo.asp?isbn=0203902750&whichpage=1&pagename=category.asp (accessed: 17 September 2014).

Calhoun, L. (2011) The end of military virtue. *Peace Review*, 23 (3): 377–386. doi: 10.1080/10402659.2011.596085.

Cantwell, H.R. (2009) Operators of air force unmanned aircraft systems: breaking paradigms. *Air and Space Power Journal*, 33 (2): 67–77.

Chamayou, G. and Lloyd, J. (2015) *Drone theory*. London: Penguin.

Cloud, D. (2011) Anatomy of an Afghan war tragedy. *Los Angeles Times*, 10 April. Available at: http://articles.latimes.com/2011/apr/10/world/la-fg-afghanistan-drone-20110410 (accessed: 17 September 2014).

Coeckelbergh, M. (2013) Drones, information technology, and distance: mapping the moral epistemology of remote fighting. *Ethics and Information Technology*, 15 (2): 87–98. doi: 10.1007/s10676–013–9313–6.

Cohn, C. (1998) Gays in the military: text and subtext. In *The 'man question' in international relations*. Boulder, CO: Westview Press. p. 129.

Cole, C., Dobbing, M. and Hailwood, A. (2010) *Convenient killing: armed drones and the 'Playstation' mentality*. [online]. Available at: http://dronewarsuk.files.wordpress.com/2010/10/conv-killing-final.pdf.

Connell, R.W. (2005) *Masculinities*. 2nd edn. Berkeley, CA: University of California Press.

Conor, L. (2004) *The spectacular modern woman: feminine visibility in the 1920s*. Bloomington: Indiana University Press.

Daggett, C. (2015) Drone disorientations: how 'unmanned' weapons queer the experience of killing in war. *International Feminist Journal of Politics*, 17 (3): 361–379. doi: 10.1080/14616742.2015.1075317.

D'Costa, B.D., (2006) Marginalized identity: new frontiers of research for IR? In Ackerly, B.A. and Stern, M. (eds) *Feminist methodologies for international relations*. Cambridge: Cambridge University Press. pp. 129–152.

Douthwaite, J.V. (2002) *The wild girl, natural man, and the monster: dangerous experiments in the Age of Enlightenment*. Chicago, IL: University of Chicago Press.

Drew, C. (2009) Drones are weapons of choice in fighting Qaeda. *New York Times*, 17 March. Available at: www.nytimes.com/2009/03/17/business/17uav.html?_r=3&hp& (accessed: 22 March 2015).

Edkins, J. (2003) *Trauma and the memory of politics*. Cambridge, UK; New York: Cambridge University Press.

Ehrhard, T.P. (2010) Air Force UAVs: the secret history [online]. Mitchell Institute for Air Power Studies. Available at: www.dtic.mil/cgi-bin/GetTRDoc?AD=ADA525674 (accessed: 18 February 2016).

Engle, K. (2009) *Seeing ghosts: 9/11 and the visual imagination*. Montreal: McGill-Queen's University Press.

Enloe, C.H. (1989) *Bananas, beaches & bases: making feminist sense of international politics*. London: Pandora.

Evangelista, M., Shue, H., and Biddle, T.D. (eds) (2014) *The American way of bombing: changing ethical and legal norms, from flying fortresses to drones*. Ithaca, NY; London: Cornell University Press.

Fingesten, P. (1959) Sight and insight: a contribution toward an iconography of the eye. *Criticism*, 1 (1): 19–31.

Freeman, C. (2015) Armchair killers: life as a drone pilot. *Telegraph*, 4 November. Available at: www.telegraph.co.uk/culture/film/film-news/11525499/good-kill-drone-pilot-true-story.html (accessed: 30 June 2015).

Freud, S., McLintock, D. and Haughton, H. (2003) *The uncanny*. London: Penguin Books.

Frosh, S. (2013) *Hauntings: psychoanalysis and ghostly transmissions*. Houndmills, Basingstoke, Hampshire; New York: Palgrave Macmillan.

Goldstein, J.S. (2001) *War and gender: how gender shapes the war system and vice versa*. Cambridge: Cambridge University Press.

Gordon, A. (2008) *Ghostly matters: haunting and the sociological imagination*. New University of Minnesota Press ed. Minneapolis, MN: University of Minnesota Press.

Grayson, K. and Mawdsley, J. (2018) Scopic regimes and the visual turn in International Relations: seeing world politics through the drone. *European Journal of International Relations*, 1–27. doi: 10.1177/1354066118781955.

Gregory, D. (2011) From a view to a kill: drones and late modern war. *Theory, Culture & Society*, 28 (7–8): 188–215.

Grossman, D. (2009) *On killing: the psychological cost of learning to kill in war and society*. Rev. edn. New York: Little, Brown and Co.

Hampson, N. (1976) *The enlightenment*. Harmondsworth; New York: Penguin.

Hansard (2015) *Armed drones*. House of Commons. Available at: www.publications.parliament.uk/pa/cm201516/cmhansrd/cm151201/halltext/151201h0001.html.

Haraway, D. (1988) Situated knowledges: the science question in feminism and the privilege of partial perspective. *Feminist Studies*, 14 (3): 575.

Harper Hart, H. (1949) The eye in symbol and symptom. *The Psycholanalytic Review*, 36 (1): 1–12.

Hasian, M.A. (2016) *Drone warfare and lawfare in a post-heroic age*. Rhetoric, law, and the humanities. Tuscaloosa, AL: The University of Alabama Press.

Hawkins, H. (2010) 'The argument of the eye'? The cultural geographies of installation art. *Cultural Geographies*, 17 (3): 321–340.

Higate, P. (2012) Foregrounding the in/visibility of military and militarised masculinities. In Baaz, Maria Eriksson and Utas, Mats (eds) *Beyond 'gender and stir': reflections on* gender and SSR in the aftermath of African conflict. Uppsala: The Nordic Africa Institute.

Holmes, K. (2015) Is today's drone tomorrow's Skynet, Terminator? CNS News [online], 5 May. Available at: www.cnsnews.com/commentary/kim-holmes/todays-drone-tomorrows-skynet-terminator (accessed: 3 July 2016).

Hurwitz, E.S. (2013) Drone pilots: 'overpaid, underworked, and bored'. *Mother Jones*, 18 June. Available at: www.motherjones.com/politics/2013/06/drone-pilots-reaper-photo-essay?ico=ushome (accessed: 29 February 2016).

Jordanova, L. (1999) Natural facts: a historical perspective on science and sexuality. In *Feminist theory and the body: a reader*. Edinburgh: Edinburgh University Press.

Kaplan, E.A. (2000) *Women and film: both sides of the camera*. Repr. London: Routledge.

Keegan, J. (2003) *Intelligence in war: knowledge of the enemy from Napoleon to Al-Qaeda*. Toronto: Key Porter Books.

Keller, E.F. (1986) Making gender visible in the pursuit of nature's secrets. In de Lauretis, T. (ed.) *Feminist studies/critical studies*. Bloomington: Indiana University Press. pp. 67–77.

Kindervater, K.H. (2016) The emergence of lethal surveillance: watching and killing in the history of drone technology. *Security Dialogue* [online]. Available at: http://sdi.sagepub.com/cgi/doi/10.1177/0967010615616011 (accessed: 5 February 2016).

Kirkpatrick, J. (2015) Drones and the martial virtue courage. *Journal of Military Ethics*, 14 (3–4): 202–219. doi: 10.1080/15027570.2015.1106744.

Kontour, K. (2012) The governmentality of battlefield space: efficiency, proficiency, and masculine performativity. *Bulletin of Science, Technology & Society*, 32 (5): 353–360.

Kronsell, A., (2006) Methods for studying silences: gender analysis in institutions of hegemonic masculinity. In Ackerly, B.A. and Stern, M. (eds) *Feminist methodologies for international relations*. Cambridge: Cambridge University Press. pp. 108–128.

Kunashakaran, S. (2016) Un(wo)manned aerial vehicles: an assessment of how unmanned aerial vehicles influence masculinity in the conflict arena. *Contemporary Security Policy*, 37 (1): 31–61.

Lee, P. (2012) Remoteness, risk and aircrew ethos. *Air Power Review*, 15 (1): 1–20.

Lee, P. (2018) *Reaper force: the inside story of Britain's drone wars*. London: John Blake Publishing Ltd.

Lyon, D. (2014) Surveillance and the eye of god. *Studies in Christian Ethics*, 27 (1): 21–32. doi: 10.1177/0953946813509334.

MacDonald, F. (2006) Geopolitics and 'the vision thing': regarding Britain and America's first nuclear missile. *Transactions of the Institute of British Geographers*, 31: 53–71.

MacDonald, F., Hughes, R. and Dodds, K. (2010) *Observant states: geopolitics and visual culture*. London: I.B. Tauris.

Martin, M.J. (2010) *Predator: the remote-control air war over Iraq and Afghanistan: a pilot's story*. Minneapolis, MN: Zenith Press.

Masters, C. (2005) Bodies of technology: cyborg soldiers and militarized masculinities. *International Feminist Journal of Politics*, 7 (1): 112–132.

Matthew, S. (2015) I'll BEE back: fleets of Terminator-style drones could have artificial brains based on honeybees. *Daily Mail* [online], 25 April. Available at: www.daily-mail.co.uk/news/article-3055152/I-ll-BEE-Fleets-Terminator-style-drones-artificial-brains-based-honeybees.html (accessed: 3 July 2016).

Parkins, I. and Karpinski, E.C. (2014) In/Visibility: absences/presence in feminist theorizing introduction: in/visibility in/of feminist theory. *Atlantis: A Woman's Studies Journal*, 36 (2): 3–8.

Porter, R. (2001) *The Enlightenment. Studies in European history.* 2nd edn. Houndmills, Basingstoke, Hampshire; New York: Palgrave.

Pratchett, T. (2014) *Thud!* Transworld Publishers.

Pryer, D.A. (2013) The rise of the machines: why increasingly 'perfect' weapons help perpetuate our wars and endanger our nation. *Military Review*, April, pp. 14–24.

Puar, J.K. (2007) *Terrorist assemblages: homonationalism in queer times.* Next wave. Durham: Duke University Press.

Rattansi, A. (2010) 'Knowing when to say when' meet Lt Col Chris Gough: killing by drone and proud of it. *CounterPunch*, 2 April. Available at: www.counterpunch.org/2010/04/02/meet-lt-col-chris-gough-killing-by-drone-and-proud-of-it/ (accessed: 22 March 2015).

Rayment, S. (2014) British drones kill hundreds of Taliban fighters in secret SAS attacks in Afghanistan. *The Mirror*, 8 February. Available at: www.mirror.co.uk/news/world-news/british-drones-kill-hundreds-taliban-3954630 (accessed: 30 September 2016).

Reynolds, E. (2015) The dark art of drone piloting: what it's like to man an unmanned craft. *Daily Telegraph*, 17 August. Available at: www.dailytelegraph.com.au/lifestyle/the-dark-art-of-drone-piloting-what-its-like-to-man-an-umanned-craft/news-story/43edfea43e7a3ad45e0149441f14250a?= (accessed: 15 March 2016).

Rose, G. (2007) *Visual methodologies: an introduction to researching with visual materials.* 2nd edn. London; Thousand Oaks, CA: SAGE.

Rowley, C. (2010) Popular culture and the politics of the visual. In *Gender Matters in Global Politics*. London: Routledge. pp. 309–325.

Royakkers, L. and van Est, R. (2010) The cubicle warrior: the marionette of digitalized warfare. *Ethics and Information Technology*, 12 (3): 289–296. doi: 10.1007/s10676-010-9240-8.

Royal Air Force (2016) Reaper MQ9A RPAS. [online]. Available at: www.raf.mod.uk/equipment/reaper.cfm (accessed: 4 June 2016).

Russia Today (2014) Fun today, gun tomorrow: how toy drones could lead to Terminator-style hunting machines. *Russia Today* [online], 12 July. Available at: www.rt.com/news/210815-drones-danger-weapons-spying/ (accessed: 3 July 2016).

Shaw, R.I.G. and Akhter, M. (2012) The unbearable humanness of drone warfare in FATA, Pakistan. *Antipode*, 44 (4): 1490–1509. doi: 10.1111/j.1467-8330.2011.00940.x.

Shepherd, L.J. (2009) Gender, violence and global politics: contemporary debates in feminist security studies. *Political Studies Review*, 7 (2): 208–219.

Shepherd, L.J. and Sjoberg, L. (2012) Trans-bodies in/of war(s): cisprivilege and contemporary security strategy. *Feminist Review*, 101: 5–23.

Sinann (2012) *Top Gun drone pilot.* Available at: www.toonpool.com/cartoons/Drone%20pilot_171950 (accessed: 3 June 2016).

Singer, P.W. (2011) *Wired for war: The robotics revolution and conflict in the 21st century.* London: Penguin.

Sjoberg, L. (2012) Toward trans-gendering international relations? *International Political Sociology*, 6 (4): 337–354.

Smith, C.M. (2016) Gaze in the military: authorial agency and cinematic spectatorship in 'drone documentaries' from Iraq. *Continuum*, 30 (1): 89–99.

Snow, E. (1989) Theorizing the male gaze: some problems. *Representations*, (25): 30–41. doi: 10.2307/2928465.

Stahl, R. (2013) What the drone saw: the cultural optics of the unmanned war. *Australian Journal of International Affairs*, 67 (5): 659–674.

Sylvester, C. (2013) *War as experience: contributions from international relations and feminist analysis. War, politics and experience.* Milton Park, Abingdon, Oxon; New York, NY: Routledge.

Taj, F. (2010) The year of the drone misinformation. *Small Wars & Insurgencies*, 21 (3): 529–535.

Thomson, K. (2009) Drone porn: the newest YouTube hit. AlterNet [online], 31 December. Available at: www.alternet.org/story/144893/drone_porn%3A_the_newest_youtube_hit (accessed: 18 February 2016).

Turse, N. and Engelhardt, T. (2012) *Terminator planet: the first history of drone warfare, 2001–2050*. Lexington, KY: Dispatch Books.

Vallor, S. (2013) *The future of military virtue: autonomous systems and the moral deskilling of the military*. Tallinn: NATO CCD COE Publications.

Van Creveld, M. (2001) *Men, women, and war*. London: Cassell & Co.

Van Loon, J. (1996) Technological sensibilities and the cyberpolitics of gender: Donna Haraway's postmodern feminism. *Innovation: The European Journal of Social Science Research*, 9 (2): 231–243. doi: 10.1080/13511610.1996.9968486.

Wall, T. and Monahan, T. (2011) Surveillance and violence from afar: the politics of drones and liminal security-scapes. *Theoretical Criminology*, 15 (3): 239–254.

Weldon, S.L. (2006) Inclusion and understanding: a collective methodology for feminist international relations. In Ackerly, B., Stern, M. and True, J. (eds) *Feminist methodologies for international relations*. Cambridge: Cambridge University Press.

Whetham, D. (2013) Killer drones: the moral ups and downs. *The RUSI Journal*, 158 (3): 22–32. doi: 10.1080/03071847.2013.807582.

Wilcox, L. (2017) Embodying algorithmic war: gender, race, and the posthuman in drone warfare. *Security Dialogue*, 48 (1): 11–28. doi: 10.1177/0967010616657947.

Williams, Alison (2011) Enabling persistent presence? performing the embodied geopolitics of the unmanned aerial vehicle assemblage. *Political Geography*, 30 (7): 381–390.

Williams, D. (ed.) (1999) *The enlightenment. Cambridge readings in the history of political thought*. Cambridge, UK; New York: Cambridge University Press.

Williams, J. (2015) Distant intimacy: space, drones, and just war. *Ethics & International Affairs*, 29 (01): 93–110.

Wood, D. (2013) Drone strikes: a candid, chilling, conversation with top US drone pilot. *Huffington Post*, 15 May. Available at: www.huffingtonpost.com/2013/05/15/drone-strikes_n_3280023.html.

Zalewski, M. (2006) Distracted reflections on the production, narration, and refusal of feminist knowledge in international relations. In Ackerly, B., Stern, M. and True, J. (eds) *Feminist methodologies for international relations*. Cambridge: Cambridge University Press.

6 Eroded souls – operational challenges to masculinity

As 75% of our shifts were night shifts, off-duty days were spent sleeping. The shift pattern was 6 days on, 3 days off (if you were lucky!). The work day did follow a pattern: wake, eat and socialise with family (length dependent on sleep), drive to work (50–60 mins). Arrive at work approximately 30–45 mins prior to crew-in time, check previously viewed flying programme still current. Check daily/weekly/monthly flying currencies and ensure all are up to date and you have read and signed for any changes. Attend crew brief (approx. 20 mins) on completion carry out any briefed pre-flight tasks. Attend 'step brief' (mini update in case anything had changed from main brief), then walk to GCS [Ground Control Station] and take over from incumbent crew. Complete briefed mission with 'comfort breaks' as required and usually a break of approx. 30 mins. mid shift. Once relieved, conduct crew debrief, fill in any relevant paperwork, check flying schedule and then return home.

(Steve)

Introduction

Steve's quote above gives an outline of his schedule at 39 Squadron at Creech Air Force Base. It provides an insight into the minutiae of Reaper operations, the small pieces that represent the underreported components of the crews' daily life. One of the things that is notable from this quote is the similarity between the lives of these crews and mine (and yours?): sleeping, eating, commuting, and spending time with family. Given that the lives of these individuals also include elements that the average life does not (surveillance, killing, dropping bombs) it is essential to consider how this blending of the normal and 'warrior-specific' components come together. What makes Reaper crews' lives so interesting is that whilst there are these similarities between their daily lives and ours, and between their daily lives and those of other members of the military, there are also components that make their experiences unique. One of these is the strange disruption of linear time in their lives, of the kind referred to as *disturbed temporality* in Chapter 2. Cycling between the home and the war, a war that is happening in a different season and different time zone, the crews are challenged by some of the markers of an apparently boundless war conducted from a distance.

Similarly, in doing so with such regularity the crews are required to shift between warrior and family identities, to be present/absent at work and absent/ present from home in such a manner that the distinctions between these zones becomes blurred, indistinct.

To understand the importance of (the disturbance of) temporality, Peter Buse and Andrew Stott's use of 'anachronism' is useful for articulating how ghosts can offer alternative readings of what has been, what is and what might be:

> ghosts are anachronism par excellence, the appearance of something in a time in which they clearly do not belong. But ghosts do not just represent reminders of the past … they very often demand something in the future.
>
> (1999, p. 14)

In Chapter 3 the ghost of the warrior indicated the way in which contemporary understandings of what it means to take part in warfare are infused with myth and history from the *past*. Similarly, some of the concerns about the use of drones (for example, the references in Chapter 5 to *The Terminator*) are framed around a fear about what the *future* of war might look like, how technologies *might* develop and what the implications could be. The present, then, is not a singular point but rather a space infused with memory and myth, with fears, hopes and imaginings. As Gordon notes, *now* is always 'laced with delight for what we lost that we never had … that it could have been and can be otherwise' (2008, p. 57). This mingling of time, the disturbance of time by ghosts, is a core component of Haunting creating what Derrida refers to as 'the spectral moment' (Derrida, 2006, p. xix).[1] According to Derrida this is 'a moment that no longer belongs in time, if one understands by this word the linking of modalized presents (past present, actual present: "now," future present)' (2006, p. xix). Through an exploration of complex personhood it is possible to see that understanding the context in which individuals function – that is, the specific mythology, the particular histories, the precise hopes and dream and fears is important to understanding the phenomena in question, in this case the interaction between the military technology of drones and gender. The precise context of this case is a British Reaper crew and this means engaging with: the social and cultural mythology and history of being part of the British Armed Forces and in particular with the legacy of previously being an all-male fighting force. As touched on in the previous chapter, it means understanding the cultural cache of being part of an active squadron in the RAF, the way the perceptions of the press and public change the behaviour (and identity) of the crews (and the ways they resist), the way the crews envision themselves in the narratives about drones, and the things that irritate them. Utilising the prism of Haunting it is possible to engage with all of these details and nuances, and indeed in this chapter I want to go further, illuminating the crews as more than *just* their jobs, situating them as part of families, friendship groups, personal and intimate relationships. In so doing I draw attention to the ways that the crews are situated in particular contexts, with particular knowledges. All of these elements come together to illuminate yet another way of understanding the crews' 'complex personhood' which has been missing in accounts of the lives of British Reaper crews.

This chapter traces three core ways in which disturbed temporality plays out in the lives of British Reaper crews. First, in their rapid cycling between the space of the home and the space of war the crews disturb the delineation between these two places. Given that the home has traditionally been coded as a feminine space and the war zone as a masculine space this has implications for the separation of the gendered dichotomy, creating a shadowy space in between the worlds where they bleed into one another. Despite commentators' arguments that the crews' space of war replicates the domestic, I argue instead that the Ground Control Station (GCS) is rather an uncanny space in which both space and time are dislocated, and therefore the identities of Reaper crews always already exceed the binary definitions of masculinity and femininity.[2]

Following this, the second way in which temporality is disturbed by the Reaper crews' experiences is through their operations. The crews might be conducting a night shift from Creech Air Force Base but that night time shift will be a day time shift over Afghanistan reflecting Derrida's claim that haunted spaces make '*Le temps est hors de ses gonds*' [time is off its hinges/is out of joint] (Derrida, 2006, p. 22).[3] The temporal disturbance is amplified by the simultaneous disturbance of space – the capacity to be apparently in *both* Lincolnshire (UK) and Raqqa (Syria), for example. The door that opens the GCS in which the crews operate serves as a boundary between two geographically non-analogous places, creating a visual *here* and *there* that does not, and cannot, exist in any real physical sense. Reaper crews operate in the strange space in which they are simultaneously at home and abroad. Operating aircraft that are flying over Iraq, Afghanistan, Libya, or Syria, British crews will be physically situated at bases in the UK or the US. Demonstrating a distinctive iteration of queer logic, the crews are simultaneously in the UK/US *and* in Iraq/Afghanistan/Syria, whilst the physical reality of that is impossible. This results in a unsettled, unsedimented category of warriors who are in and (in) between spaces, warriors who are here but not here, there but not there.[4]

The third and final way in which this chapter engages with Haunting's concern with disturbed temporality is through looking at the implications of the spatial and temporal dislocation of the first two sections. I divide these implications again into two further sections. The first explores the way that cycling rapidly between the space and time of war and that of the home, the crews have little opportunity to adjust their mind-set. Drawing on research into the importance of decompression and reflecting back on the spectres of trauma raised in the previous two chapters, this section explores the way that crews need to find ways to draw lines in the liminal spaces they inhabit. Without being able to do so, I argue, the crews risk enabling the bleeding of the war mind-set and behaviours into the domestic. The second implication of temporal dislocation is the crews' experience of chronic fatigue. I explore the ways in which fatigue serves to both masculinise and feminise the crews, drawing on the sensuous knowledge of symbolism to counter unfavourable comparisons between the drone machinery and the human crews.

The warrior (not) abroad

Part of what makes war different and distinctly separate from the home and the domestic has historically been the physical and geographical distance between the two spheres. The warriors outlined in Chapter 3, be that Achilles or Odysseus, St George, the Knights Templar or the 'Band-of-Brothers' were all required to go *away* from their home spaces to battlefields that were geographically discrete, even remote. This created temporal-spatial lines that informed the distinction between the home and the war front, the domestic/private and the public, and informed statements like 'this is no place for a woman' in relation to the battlefield. War then became not just something men *did* but also a place where they *were* (and a place where women were *not*), reinforcing the binaries between the masculine and feminine. Similarly, the domestic became a place where war was not, it was a safe haven (ideally), and the protection of that *space* (as much as the individuals within it) became an important marker of the warrior-identity.

In 'going away' to war, the warrior was provided with the space in which to demonstrate markers of masculinity such as self-sufficiency, and courage. The experience of 'danger and deprivation in a distant place' is part of what constructs the warrior's identity, and draws together the individuals in units in 'comradeship' (Allsep, 2013, p. 387). In traditional narratives and mythology (and often real-life) the arena in which war takes place is dirty, confusing, chaotic, and dangerous, whereas the domestic/home space is (ideally, or comparatively) clean, calm and safe.[5] Similarly, as Braudy notes 'In the basic equation of traditional warfare, honour was indirectly proportional to distance' (2005, p. 385), thereby implying that those warriors who stay close to the home space are feminised by their proximity, and considered less-masculine than those fighting a long way away. In specific relation to the British military, this can be understood in connection to colonial endeavours. These narratives speak to the Chivalrous Gentleman iteration of the warrior, with warriors going away to fight 'primitive', 'exotic', and 'barbaric' peoples. The aim of doing so was not only conquest (although this was certainly important) but arguments were made to claim that these warriors brought those 'natives' the 'gifts' of Christianity and European 'civilisation'. In order to undertake these kinds of 'civilising' missions, warriors were required to show specific kinds of bravery (as outlined in Chapter 3), and preparedness to withstand the horrors of the unknown.[6] Similarly, 'going away' enables the warrior to 'return' triumphant, coming back to the domestic space having conquered whatever evil, and having done his job as 'protector' of the (domestic) space.

Given the importance of the space of war in delineating the warrior, what does this mean for members of the military whose experience of war does not take place on the battlefield (as traditionally understood)? Utilising Haunting means paying attention to the situatedness of the Reaper crews both temporally and spatially, including the space that they occupy during operations. Whilst they do not 'go to war' in the way we traditionally understand it, they do still have a specific space in which they conduct operations. The GCS provide the space in which the crews are simultaneously absent and present, here and there, then and now. Whilst the GCS are physically located in either the UK or USA the operations that the crews conduct are, as previously noted, based somewhere else (previously Afghanistan and Libya, and currently Iraq and Syria). The GCS are described as

> the size of a metal shipping container with a door on one end. Inside, a thin carpet covers the floor and a bank of monitors with two chairs sits at one end. Several air conditioners hum, keeping the electronics in the GCS cool. The lights are dim so the pilot and sensor can see the monitors.
>
> (Maurer, 2015)

The darkness of the GCS is commented on in a number of reports. Described variously as 'dark bunkers' 'dim, chilly, damp trailer[s]', and 'cool dark room[s]' these accounts add to the ghostly and clandestine atmosphere associated with the world of the Reaper operator.[7] It is important to note here, that these journalistic accounts offer a stark contract to the descriptions from Peter Lee's ethnographic research, which reveals something significantly less menacing, despite the relatively low-lighting:

> This is definitely not a secret shiny drone lair. If the interior designer is aiming for jumble sale chic ... then the look is a triumph. The ambience is completed by the lingering scent of microwaved curry ... I am fascinated by the fact that the walls, floor and ceiling are all carpeted.
>
> (Lee, 2018, pp. 21, 29)

The phrase 'jumble sale chic' does not conjure up foreboding sentiments of *Terminator* futures, but rather hints again to the domestic, even as it sits cheek-by-jowl with computer screen after computer screen, long thick snakes of cabling, and the paraphernalia of the controls for the aircraft itself. Indeed, even though the journalists' descriptions of being 'cool and dark' indicate a somewhat unfriendly environment, these descriptions remain a world away from the battle-field as chaotic, dirty, and dangerous.

Within the GCS, the three-person crew (pilot, sensor operator and intelligence analyst) are situated in front of a huge bank of computers (Mead, 2014).[8] The screens show everything from live video feeds to secured chat room conversations, to maps, to technical readings from the aircraft. As outlined in the chapters on strikes and surveillance, the Reaper has a considerable battery of technical capabilities, all of which need to be linked in to the crews in the GCS. As a result, crews are required to utilise a range of different computer screens and read outs in addition to maintaining voice (including radio) and text communications with individuals in command, individuals on the ground, and others. In addition to the computer screens, 'The pilot's station resemble[s] a traditional aeroplane's cabin or cockpit that used a standard flight stick, rudders, trim tabs, throttle and other associated controls' approximating as much as possible the 'standard' operating machinery of any other aeroplane (Martin, 2010, p. 19).

What makes the situation of the GCS so interesting is that it is haunted by narratives comparing it to a glorified office environment or a child's bedroom (or indeed the chaos of the jumble sale). As Cara Daggett notes, 'It is difficult to find a description of drone warfare that does not include … that drone operators are surrounded by the accoutrements of the office worker, sitting in ergonomic chairs, drinking coffee and eating junk food' (Daggett, 2015, p. 367). The distinction, in those narratives, between the war space of the Reaper crews and the war space of 'real' soldiers is used as a means of indicating the Reaper crews' lower status. The GCS is an environment in which it is possible to engage in 'push button' warfare, without physical strength, without valour, without bravery (Mayer, 2009). The disturbance of the space of war, so that it is overlaid with some of the accessories of the domestic (the perennial references to the ergonomic chair), creates a space that jars. The war space is rendered uncanny in comparison to our traditional understandings of what that space is like, the familiar narratives from history and myth are jarred by a space that is neither battlefield, nor office space, nor home space.

In addition to rendering the battlefield uncanny, the GCS also serves to disturb the crews' experiences of time and geography. The GCS are set up in shipping containers because this makes it easy to transport them and to plan appropriate amounts of space for them on various bases. In addition, the 'closed' nature of the GCS helps to separate the crews from the world at home from the world at war: an absence from the home (even though the crews are close to home) and a presence in war (even though the crews are not, in fact, actually where the war *is*). In this space day-time can also be night-time, and Las Vegas sunshine can also be low hanging clouds over the Hindu Kush. Justin noted

> we worked in three different time-zones. The army like to do everything in local time to where they are at the time. Aviation is always done in 'Zulu time' [Universal Coordinated Time (UTC) based on Greenwich Mean Time], and then there was obviously local time where we were in Creech, in which you did your shifts, and lived your life.
>
> (Cole, 2017)

Additionally, interviewee Geoff noted that this dislocation created not just mental confusion but also bureaucratic, noting 'it was night-time Creech, which is daytime Afghanistan, where do you put that in your log book?'[9] Was it a night flight or a day flight? Did it matter more what time it was where the crew were situated or the time where the aircraft was situated?

My aim in sketching all this detail out is to situate the crews in the *place* that represents their curious situation in war. Despite claims about 'armchair killers', 'Dilbert at war', 'the cubicle warrior' and so on, the GCS is very different from the usual office space (or domestic space).[10] And yet narratives that connect the crews' 'space of war' with domesticity continue to haunt the reports about

Reaper crews, influencing public and military perceptions of the crews. Would the public feel better, I wonder, if the crews' chairs were less comfortable, less ergonomic? Would that make us feel that they were enduring the discomforts we expect of warfare and would we, therefore, be ethically more comfortable with the role they play?

To reinforce the liminal space between war and home that the doorway of the GCS represents, the crews rely on 'sensuous knowledge', immersing themselves as much as possible within the world of the war that the GCS seeks to encapsulate. The doors to this area are sometimes marked with signs like 'Welcome to Iraq' or 'You are now entering Afghanistan' to illustrate the transition to the battlefield 'space'.[11] As described by one pilot: 'When I am in the GCS, I *am* in Afghanistan' the door acting like a time and space portal to the space and time of war (Lee, 2018, p. 170). The absorption within the GCS is such that many individuals note a sense of confusion and/or amazement when they finish their shifts and step back out into Creech/Waddington:

> For the crews when they get into the cockpit they are, heart and mind and soul, at the other end of the sensors and the displays, and they're in the mission, in that conflict. The really really bizarre bit is stepping back out and realising you are in Lincolnshire.
>
> (Killeen, 2015)

Or similarly as Col Black noted

> my particular portion of that mission was over and all of a sudden the door opens on the box and the next pilot and, and sensor operator walk in and 'Oh my God, I'm not, I'm not in Afghanistan [laughing], I'm still in Las Vegas' and it's extremely strange, it's very strange.
>
> (Black, 2013)

As one of my interviewees noted after an engine failure, and the crashing of the Reaper, it took the crew a few seconds to realise that they were not physically in the aircraft, and that they were not hurt by its fall.[12] Even in the usual course of flying the crews respond physically as if there *are* present in the aircraft:

> I can tell he has gone into a left turn. In an aircraft cockpit in flight, gravity and centrifugal forces combine to cause the crew to lean into the turn. Despite this particular cockpit being a shipping container in the middle of the Nevada desert, the pilot still leans left into the turn.
>
> (Lee, 2018, p. 36)

What this illustrates is the power of sensuous knowledges: the crews *know* at a rational level that they are not physically present in the Reaper, and yet they *feel* that they are. The shock of the crash landing resulted in the same spike of adrenaline that the crew might have experienced in a crash in which they were physically present, the sensation of dislocation and *un*realness when stepping back into the geographically *real* environment of Waddington or Creech exemplified some similar effects. The concept of disorientation speaks to the idea of ghosts, reflecting the idea of the familiar made strange and of the appearance of the 'unreal'.[13] The sensation of being 'transported' into the theatre distorts time and space un-naturally (supernaturally?), creating a neither-here-nor-there-ness which means that the individuals are capable of 'los[ing] sense of the fact that you are sitting in Nevada' (Black, 2013). Importantly, and this is often overlooked, it is not just once a day that the crews have to remember that they are *not* physically present in Afghanistan or Iraq (or wherever the Reaper is) but in the UK or US, but after each part of their shift – so in the British case, about every three hours.[14]

These sensations are not available to the individuals who commentate on the environments in which the crews function. Without the use of Haunting it is difficult to capture the nuance (and queer logic(s)) of the crews *simultaneously* being and not being present in the war. All that is seen is the difference between the experience of those *geographically* within the war space (understood as a chaotic, smelly, dangerous, traditional, and above all *masculine* battlefield) and the comparative cleanliness and quiet of the Reaper crews' operating environment that marks them out as feminised in comparison.[15] And this dislocation can be, as one individual succinctly put it 'a mind fuck' (Lee, 2018, p. 55)

Bleeding boundaries: 'I killed six Taliban today in Pakistan, now I'm off home to barbecue'

Understanding the crews as operating in a liminal zone where traditional barriers become translucent and ghostly provides an important space for talking about the peculiarity of the work/life balance of these individuals. The uncanny, in the battlefield rendered strange by the GCS, is also evident in the disturbed temporality between the home and the war and the experience of travelling rapidly between the two. The distance between the war and home, and the time taken to travel between the two, not only cemented the markers that defined the difference between them, but also provided an important personal space in which the warrior could move psychologically and emotionally between the necessary mind-sets. In contemporary discussions, training provides the preparation for the move from the home to the war, and the process of reversing that training to move back from war to home is referred to as decompression.[16] There is extensive research that demonstrates the importance of decompression in mitigating the potential trauma of reintegrating military personnel into civilian society.[17] As one interviewee summarised:

> When infantry are deployed they stick you in Cyprus for a couple of weeks for decompression … [without decompression] testosterone fuelled guys were coming home leading to domestic abuse because they couldn't switch that off.[18]

The importance of decompression in preventing mental health problems in servicemen/women has also been acknowledged by Members of Parliament in the House of Commons. As such, in a 2008 report on Medical Care for the Armed Forces it was noted that decompression 'should be an integral part of the procedures for all personnel returning from operational tours' (House of Commons, 2008, p. 29).

The quote above indicates the way in which decompression is important not just for the personnel themselves but also for the safety and well-being of their family and friends. The individual who noted during the strike between himself and the men on the ground, that 'They were just men like me' then went on to state 'And then I'd come home after the shift and I'd be handed a baby. Thirty-eight minutes was my drive home' (Lee, 2018, p. 224). As this indicates, for British Reaper crews decompression, in any meaningful way, is not available to them. For these individuals the longest period of decompression is a 50–60-minute drive back to the suburbs of Las Vegas from Creech Air Force Base. Or if, as the individual above, they live on base at Waddington they have even less time to transition from 'war' to 'domestic' spaces, with all the accompanying shift in mind-set. With the development in the use of smart phones this can be particularly jarring:

> When that [shot] was finished I came straight upstairs and there, on my phone, my wife had sent me a video of our son walking around and playing with his new toys. Mentally, I had moved instantly from an operational video feed and a shooting scenario to a family video. She … asked, 'What do you fancy for dinner?' I thought *I don't know, I don't know. I can't think about that.*
>
> (Lee, 2018, p. 221)

This speed of transition is important for considering the strangeness of the switch between the home and war, an experience that is jarring without adequate time to make sense of the shift psychologically.[19]

Where decompression is not available to crews the ghosts of recent or previous operations can surface in the space of the home. As a result, crews had to find their own ways of dealing with the warrior/home switch. Interviewees gave me the following examples:

> you might go to the break room to calm down for half an hour and then get in your car and drive home … you are not in the war any more....
>
> (David)

> It's hard … [but you] find that opportunity to decompress.
>
> (Freddie)

> You'd creep home … you['d] have a cup of tea and a bit of a think … and then [go] to bed.
>
> (Robert)

Different people used different coping mechanisms for compartmentalising ... some people found it really difficult ... some people found it really easy ... the majority felt a degree of stress with it ... sometimes I would need to go home and decompress....

(Peter)

ignore wife and kids, drink beer.

(Anonymous)[20]

These quotes indicate some of the small and particular ways in which the crews attempt to deal with the dislocation of living at home *and* at war, of being in the UK/USA *and* operating in Afghanistan/Libya/Syria. My interviewees were not unusual in their claims; as similar comments are reported by others:

I'll spend an hour so in the gym or go on a run ... I create a buffer zone in my mind to separate my civilian life from my work life.

(RAF Squadron Leader cited in Peterson, 2016)

It's lonely. I can't tell my partner what my day at work was like. This morning I'm out walking my dog, through the soaked wet fields, no one for miles around. I find myself sat down talking to my dog for comfort. Her unconditional love keeps me strong. It's a weird life.

(Lee, 2018, p. 211)

It takes some self-conditioning.... You have to get good at compartmental-izing.... If you go home and have dinner with your family, you don't want to think of something that happened earlier in the day.... You think about it later.

(Martin cited in Mulrine, 2012)

[US Sensor Operator] Sparkle came home to her dog, eager to head out to the park ... Sparkle likes it because no one is in the military ... 'It's great to sit and just share a love of dogs,' she said. 'They make me happy to protect a world where we can spend lavish amounts of money on recreation places for our animals. The women I met there were strong and independent ... and [the dogs] give us some subject to talk about other than work'.

(Maurer, 2015)

These quotes and the ones from my own interviewees above, indicate the way that the crews attempt to keep the ghosts at bay by compartmentalising. Reflecting back on the visual trauma addressed in the previous chapter, it is evident that in some cases the ghosts of the war seep through into the spaces of the domestic. The mental video loop of the deaths of friendly forces or of innocents, the mangled limbs of the battle damage assessments; all threaten to haunt the crews' day-to-day lives in ways that go beyond affecting military operations.

One of the worries about the temporal and spatial dislocation of Reaper crews, and the attendant lack of decompression, is the potential for the intrusion of war behaviours and thinking patterns into the domestic space. The intrusion of the masculine into the feminine, the *masculinisation* of the domestic. Whilst I am certainly not the first to discuss the 'militarisation of the everyday', the gendered implications are particularly complicated when looking at the day-to-day cycling between war and home experienced by Reaper crews.[21] A number of the individuals that I spoke to had children, and still more had partners/wives/boyfriends/girlfriends. The different identities and roles that crew members have to navigate creates tension as the two worlds refuse to remain in the separate compartments but threaten to spill over into each other. The video of a children learning to walk or playing with his new toys, where does this fit into the GCS after a kill shot? This reflects Gloria Anzaldúa's (2012) identification of borderlands as places and spaces of anger, insecurity, and violence. In the film *Good Kill*, the lead character (Tommy) is haunted by his work and unable to cope with the disjunction between his home life and life in the skies above Afghanistan. As a result he is unable to interact normally with his wife and children, seeks solace at the bottom of a bottle of vodka and lashes out violently at his partner (*Good Kill*, 2015). Whilst this is a fictionalised account, and interviewees described the characters as caricatures, they also noted that Tommy's responses are 'accurate but accentuated' and it was 'fairly accurate ... from the fatigue point of view and getting wound up' without decompression.[22] In a recent debate in the House of Commons, Members of Parliament debating the use of armed drones included the impact of a lack of decompression in their discussions, noting:

> The preparation and processes that they undergo are exactly the same as those required for flying a conventional aircraft, and once the door to the workspace is closed the pilots report that it puts them psychologically in that airspace, with all the emotions and thought processes being exactly the same as on manned planes ... [But that] One RAF crew member is reported as saying that the potential for psychological and emotional impact on drone operators was 'far greater than it ever was with a manned cockpit.'
>
> (MP Kirsten Oswald quoted in Hansard, 2015b, Column 56WH)

As a result of the potential for 'psychological and emotional impact' and in recognition of the 'unique nature of such operations', Penny Mordaunt, the then Minister for Armed Forces, stressed the access to mental health care for Reaper crews (quoted in Hansard, 2015a). She said,

> [W]e have embedded TRIM – trauma risk management – providers in RAF Reaper squadrons. As hon. Members know, TRIM provides a model of peer group mentoring and support for use in the aftermath of traumatic events.
>
> (Mordaunt quoted in Hansard, 2015a, Column 70WH)

Similar points were also made in another debate on the use of drones in conflict.[23] Interview data suggests that psychological support services are being made use of, and there is a growing wider acceptance that drone crews are not immune to the violence that they witness (as per the previous chapter). What has yet to be fully realised is the effect that this will have on the traditional markers of stoic, rational military masculinity. In addition, the lack of institutional experience of this phenomenon (of drone crew shift work) means that crew members are (to some extent) having to find their own ways of dealing with the complexities of switching between war and home on a daily basis.

Soul searching

In this section I return to the division between the home/war spaces to consider how this division is also disturbed by the crews' experiences of cycling between the two and the impact of chronic fatigue on relationships with their families. Focusing through the prism of disturbed temporality, it is possible to explore how flying night shifts that are simultaneously day shifts, commuting to and from work in UK/USA but flying in Afghanistan/Iraq/Libya/Syria, has effects on the crews' personal and social lives in ways that have implications for our narratives of masculinity and femininity in war. As outlined at the start of this chapter, Haunting is concerned with the complexity of individual life and that means acknowledging that the crews are more than *just* their jobs, but complex human beings embedded in a multiplicity of different kinds of relationships. As Chapter 3 laid out, the ghostly warrior develops part of his status from his domestic interactions, particularly noted in the Husband–Protector iteration. This iteration hinted at what the Reaper crews' experiences have highlighted, that the separation between home space and war space, whilst historically/mythologically distinctive, is not as clear cut as is sometimes portrayed. This section then traces how the separation is disturbed by the crews' experiences of fatigue and the impact that this has on their identities not just as warriors but also as family members.

In the Husband–Protector iteration of the warrior, with this individual at the head of the household, he is responsible for protecting his family from the aggression of others and laying down his life if necessary. By applying the framework of Haunting it is possible to direct attention to the tension that exists at the very centre of this warrior identity. The warrior mind-set requires the capacity for violence, even brutality, in a way that requires the individual to deny or mitigate emotional connections with the enemy. The husband/father must be capable of precisely the opposite, demonstrating benign, even gentle, caring forms of masculinity based on an emotional connection with his family members.[24] The need to navigate these dual identities is something that crops up in memoirs and interviews with military personnel involved in warfare in a range of different ways and is therefore not unique to Reaper crews. However, the daily tempo of Reaper operations, and the rapidity of cycling between the two *is* unusual and this creates its own particular pressures. One US pilot spoke about the pressures, the problem is you can't talk about your work, what you have seen, or what you have done, because of security…. Pretty soon, spouses don't understand why, and the friction really begins. In many ways, I wanted to tell my wife everything, but knew that I couldn't, so we mainly focused on how her day went. Needless to say … that led to a lot of pent-up stress.

(Cited in Chow, 2013)

The impact on marriages and relationships is something that crops up a lot in discussions of drone crew stresses.[25]

My interviewees indicated that the biggest issues developed for couples where the wives were 'non-military' or had little experience of military life. In these instances, the family unit became haunted by the narratives of the traditional 'heteropatriarchal family/households', reflecting the historical dominant Western paradigm of the family in which 'men were the breadwinners and heads of the household while women were full-time wives and mothers' (Peterson, 2014a, p. 605; Niva, 1998, p. 110). As one interviewee noted, this caused a lot of marital discord because the wives (of crews based at Creech) 'had to fit into mid-twentieth century officers' wives' roles'.[26] This reflects a temporal disturbance of a different kind – a shift backwards in history in terms of gender and the roles that individuals can take on in their intimate relationships set against the introduction of one of the most cutting-edge technologies of warfare.[27] For some of the crews this challenged the roles that they had previously taken up with their partners. As one interviewee noted, the job 'has to come first' and as a result, 'a lot of other people do get hassle for that … particularly [within] 39 squadron' where the wives and children have been uprooted from their homes in the UK and deposited into the suburbs of Las Vegas. And those who fought to maintain their, perhaps, more equitable relationships endured the stress of 'rushing home after stressful shifts to do the housework' or worrying about sick children.[28]

As well as relationships with spouses, complicated relationships with children emerge through the disturbance of temporality. One lieutenant colonel of the USAF described the disconnect between work life and home life with children:

Every decision you made was … somebody living, saving somebody or somebody dying. And you walk into your house and you're trying to figure out whether your daughter is going to wear a blue tutu or a pink tutu and the disconnection is astounding.

(Black, 2013)

Similarly, 'Oz' states

the weirdest thing for me – with my background [as a fast-jet pilot] – is the concept of getting up in the morning, driving my kids to school and killing people. That does take a bit of getting used to.

(Cited in Blackhurst, 2012)

Unlike spouses, children cannot necessarily understand why mummy or daddy is particularly distant this evening. The space the children occupy, the home space, is traditionally sectioned off from the concerns of war. One of the 'benefits' of the posting to Reaper squadrons is that the individuals are not away from their families for extended periods of time. They can, depending on shifts, perhaps avoid missing out on the some of the milestones and important (as well as day-to-day events) that other deployed military personnel miss out on. However, the invasion of the home space with the concerns of war is neatly summed up by one of my interviewees who recalled checking the weather in Kandahar whilst he was watching the TV with his children, the cartoons of Nickelodeon set jarringly against the wind speeds above the theatre of battle.[29]

Dreams in infrared[30]

Partly as a result of the blurring of the home/war boundary, and the concomitant blurring of the masculine/feminine binary creating a 'disorientation', British Reaper crews are experiencing high levels of fatigue (Daggett, 2015). This fatigue, linked as it is with the limitations of the body, haunts the Reaper crews with suggestions of feminisation: The body being coded as feminine and the mind as masculine.[31] Additionally, the warrior is traditionally constructed as overcoming fatigue and pain to achieve tremendous and unusual feats. Thus, the crews' experiences of fatigue then serve to situate them as less masculine and as failing in the performance of warrior behaviour. I argue in the following section that this feminisation functions in opposition to the steely, indefatigable body of the drone, which is then imbued with the status of the masculine.

Reflecting the curious presence/absence of ghosts, the drone situates the crews at a distance from the traditional theatre of war whilst their bodies still are in the war. Distanced yes, but disembodied no. For example, having a three-person crew that enables a 'three-way pee break' highlights the need for the human bodies (and the organisation of those bodies) to shape themselves around the drone's lack of need.[32] The eyes of crews, unlike the video-eyes of the Reaper, may droop with tiredness; their concentration may waver in boredom or as a result of fatigue. In comparison, the drone is *always* 'watching', streaming data and information.

Dealing with changing shifts disrupts the circadian rhythms of the crews. The temporal disturbance of night shifts that are also *simultaneously* day shifts is woven into a day-to-day pattern of shift rotation. Therefore, the experiential dislocation outlined in the previous sections is also woven through with a *biological* dislocation which comes from cycling through shifts that do not necessarily fit to the body's natural rhythms. The crews usually do a rotation of six days on and three days off and the shifts vary between eight and 12 hours in length, broken into three hour sections.[33] Individuals are assigned to various different shifts so that they might need to switch between day and night shifts.[34] Most of my interviewees agreed that the most fatiguing shifts were the night shifts and in recent studies of US Predator/Reaper crews, fatigue was found to be the primary reason for crew burnout.[35]

Robert claimed that he initially thought it was a 'sweet deal' to move to Las Vegas. However, after a second month of night shifts he was wondering whether he had made the wrong decision.[36] The effect of disrupting circadian rhythms is reflected in other interviewees' attempts to manage the discomfort that results. For example, David relied on 'sports drinks to wake up [and] sleeping pills to sleep'.[37] Similarly, an anonymous interviewee claimed that he had 'spent a year taking herbal sleeping pills because [he] could not sleep'.[38] Both British and American Reaper crews struggle with the long, straight drive back from Creech Air Force Base to their homes in and around Las Vegas, with many reportedly nodding off at the wheel. One wife of a Reaper operator recalled:

> It wasn't unheard of in those early days at Creech for him to say, 'I had to pull over to the side of the road and sleep in the car for an hour on the way home because I was nodding off. I nodded off three times.' On two occasions he told me that he'd been woken up because he banged the kerb on the side of the road because of the level of exhaustion he was experiencing.
>
> (Lee, 2018, p. 264)

As a result, perhaps, there is a room of beds made up for those who feel too tired to drive home safely (Lee, 2018).

To mitigate the effects of night shifts, numerous interviewees spoke of needing to sleep through all three of their days 'off'.[39] Needing to sleep through all of their days off prevents crews from spending the time doing necessary life administration. In so doing they are prevented from socialising or spending time with their families, arguably creating a backlog of issues and requirements that intrude into the days when they are working: the boundaries of war and home bleeding into one another in reverse. The crews' experience of fatigue is set against the spectral shape of the Reaper drone itself, drawing attention to the bodily limitations of the human crews, a comparison which has both feminising *and* masculinising implications for Reaper crews' identities as warriors.

The Reaper crews are haunted by the unfair comparison between the capabilities of their physical bodies and those of the cyborgian 'drone'. The drone indicates that the British Army tagline of 'be the best' no longer applies to human beings, being the 'best' for these kinds of operations means being more-than human or (at least partly) machine (Be The Best, 2014). In this context, where Cristina Masters argues that the cyborg demonstrates the 'techno-scientific and masculinist discourses of power', it also seems that these discourses reveal the limitations of the fleshy male body (2010, p. 2). In this sense, Allsep's (2013) concerns that hegemonic masculinity is challenged by the capacities of technologies which exceed those of the warrior appears to apply. As such, Reaper crews are haunted by the cyborg that they cannot be, functioning as they do on the boundary between wakefulness and sleep. What the interviews with the Reaper crews has revealed is that, far from it only being women, children and targeted 'Others' who are limited by their bodies, so too are the soldiers, the warriors. Even those men whose 'bodies are often aligned with technology ... [with their] body as instrument or weapon' (Pettman, 1997, p. 95). Therefore, Reaper crews' capitulation to fatigue constructs them on the feminine side of the binary, the ghost of the cyborg revealing the comparative masculinity of the drone machine.

However, looking at the situation again and embracing queer logic, it is also possible to argue that the crews demonstrate their masculinity by functioning through their fatigue. After all, despite their tiredness, despite complaints about doing six months of night shifts, despite having to function in worlds that are always here and simultaneously there, and despite the rapidity of cycling between the home and the war, the crews continued to do their job(s). I was interested in the language that Geoff used to describe the situation, as he noted that the crews would 'take the pain to stag on'.[40] This phrasing is interesting for an analysis of masculinity in the lives of Reaper crews for two reasons. First, the reference to enduring pain is reflective of the warrior trait of stoicism and the capacity for endurance.[41] As Braudy notes,

> In the vast variety of initiation rituals, the implicit question always seems to be, 'Can the candidate endure pain?' This draws a connection between the initiation of the warrior and the initiation of the individual into manhood 'in which the candidate must move from the unripened sexual identity of childhood into a world of adult gender difference ...' through the endurance of physical discomfort.
>
> (2005, p. 19)

As noted in Chapter 4, Reaper crews have been critiqued for (and perhaps haunted by) their lack of physical risk (which has led to a questioning of their warrior status). However, to refer to the crews 'taking the pain' runs counter to this argument and draws attention to the different kinds of pain which a warrior might be required or be able to 'take on' and endure as a means of demonstrating their masculinity.

The second reason for an interest in Geoff's phrasing is the reference to the 'stag'. Within discourses of masculinity, the stag has a range of different, but ultimately reinforcing, meanings. As noted in Chapter 2, symbols are important to my exploration of the uncanny, representing a form of sensuous knowledge and the symbol of the stag is woven through with meanings related to military masculinity and the lives of Reaper crews. To 'go stag' is to go it alone, to prove one's independence, a trait Kaurin (2014) identifies as a core component of being a warrior. Historically the stag was the ultimate hunting prize and the capacity to kill one reflected both prowess with a weapon on horseback and the ability to endure the physical demands of an often long and arduous hunt (Mangan, 2015; MacKenzie, 1997). Hunting stags was also associated with high ranks, the domain of the nobility, and therefore connected with narratives of the knights and chivalrous masculinity (Richardson, 2012; Ehrenreich, 1998). Perhaps, at the most straightforward level, the deer is an animal that displays strong differences between the males and females of the species. We also use the biological sex different between deer to reinforce gendered descriptions of men and women. For example, we refer to individuals as being 'doe-eyed' when gentle, and as Sarah Gothie notes, stag antlers

> are first and foremost rugged, straightforward masculine symbols. The most impressive 'racks' come from male animals and have been displayed historically by male hunters as trophies – in a variety of cultural and class contexts.
>
> (2015, p. 15)

Without wanting to read too much into a short comment, I found it useful to consider some of this symbolism as haunting the Reaper crews, influencing how they envisioned themselves as performing masculinity. 'Taking the pain' and 'stag-ing on' reinforces not only the masculine warrior identity of the crews through the ways suggested above, but links to the RAF conceptualisation of 'professionalism' and masculinity. Professionalism is one word which comes up again and again in interviews with Reaper crews, and particularly in interviews with *British* Reaper crews. For example, in one interview with Adrian Chiles from the BBC (Killeen, 2015), three Reaper crews and the Wing Commander made 18 references to professionalism in less than an hour. By 'stag-ing on' crews pushed themselves to the limits of fatigue in order to support fellow forces on the ground or to meet important mission objectives, demonstrating their professional commitment and their ability to put others' needs before their own. Despite the uncomfortable dislocating experience of temporal disturbance which haunts the lives of the crews, an experience that situates them both here and there, and then and now, my interviewees distanced themselves from the emotional language of the press by calling on professionalism as a marker of what made them both ethical and distinctive. Tied to ideas of rationality, reasonableness, and technological mastery, professionalism acts as the signifier for this constellation of masculine traits.[42]

Conclusion

The lives of British Reaper crews are permeated with the different ways in which temporal disturbance operates. Engaging with questions of presence/absence and home/abroad, or home/war; this chapter has illustrated how Haunting provides a means of taking this kind of disturbance seriously. Acknowledging the way in which the binaries associated with temporal disturbance are gendered, I have illuminated how the blurring of these binaries serves to disturb not just time, but also Reaper crew identities. Focusing on the differently gendered spaces of home and war, the history and myth of the warrior 'going away' to war (creating a space of war and a space which belonged to men) is destabilised by the way in which the Reaper crews commute to and from the home on a daily basis. Similarly, disturbed temporality emerges as an important theme as the crews' operations made time seem out of joint/off its hinges (Derrida, 2006), amplified by the experience of not only flying night shifts that are *simultaneously* day shifts, but also existing in a borderland between US/UK and Afghanistan/Libya/Syria which is geographically impossible. All this temporal disturbance in the daily-lives of the Reaper crews has implications for how they interact with their friends and families. The relative lack of decompression available to the crews raises the spectre of the violent masculinity of the war bleeding into the other-wise protected zone of the domestic. As such, situating this domestic space as populated by wives/partners and children represents a disturbance of temporality that could have material and violent effects on military families. Additionally, the experience of chronic fatigue initially appears to make it difficult for the crews to replicate the stoicism of the warrior, destabilising military masculinity. However, 'sensuous knowledges' provide an insight into alternative readings of the experience of fatigue in ways which actually reinforce military masculinity.

Temporal disturbance is also implicated in the crews' experiences of chronic fatigue. The cycling between day shifts that are *simultaneously* night shifts, is dis-rupting the circadian rhythms of the crews, and causing dislocation which is experi-enced bodily through disturbed sleeping patterns. The ghostly comparator of the cyborgian drone, which is not fatigued, serves to feminise the Reaper crews by con-necting them with the concerns of the body. Therefore, whilst some commentators have argued that, in piloting drones (viewed as cyborgs) the crews demonstrate hypermasculinity; I have argued in this chapter that the reverse is also true.[43] Endur-ing the pain of fatigue operates as a means of challenging the crews' haunting by the cyborg. Through acknowledging their bodily fatigue and *overcoming it*, the crews utilise the temporal disturbance at the heart of their operational experiences to demonstrate their military masculinity, embodying the warrior capacity to prevail.

Notes

1 (See also Kenway, 2006; Coddington, 2011).
2 (Mayer, 2009; *The Economist*, 2014; Freeman, 2015).
3 This is a result of 11.5 hours of time difference between the time zone in Nevada and that in Afghanistan.
4 (inter alia Daggett, 2015).
5 (Elshtain, 1995; Horn, 2010).
6 (Dawson, 1994).
7 Would these narratives survive as easily, I wonder, if Reaper crews operated in bright, well ventilated, sun-lit rooms? (Skaar, 2014; Mead, 2014; Chow, 2013).
8 In the US iteration the pilot and sensor operator sit in the GCS, and the intelligence analysts sit together in a separate section. The British configuration emerged because British intelligence analysts could not sit with the American intelligence analysts, but crew members now talk about this triad as an integral part of what makes the British use of Reaper coherent and useful. In addition, Geoff pointed out that the three-man crew enabled more 'pee breaks' than in a two-man crew – an important consideration in manning decisions!
9 Interview with 'Geoff'.
10 (Freeman, 2015; *The Economist*, 2014; Royakkers and van Est, 2010).
11 Interview with 'Peter'.
12 Interview with 'Peter'.
13 (Gordon, 2008; Derrida, 2006).
14 (Lee, 2018).
15 (Manjikian, 2014; Kunashakaran, 2016).

16 (Hughes *et al.*, 2008; Garber and Zamorski, 2012; Fertout *et al.*, 2012).
17 (see for example Zamorski *et al.*, 2012; Garber and Zamorski, 2012; Green *et al.*, 2010; Walker, 2010).
18 Interview with British military intelligence analyst part of Royal Air Force 39 squadron based at Creech, USA
19 Some existing RAF pilots point out that it is not entirely new to be able to conduct a lethal strike and then be home for breakfast with the kids – this was something some fighter pilots experienced during operations in Kosovo (Cole, 2018). But what is new, and currently unique to Reaper crews, is the regularity and length of time for which this is a reality they must wrestle with. For the pilots over Kosovo their tour would have been months. For the Reaper crews it is years.
20 Interview with 'David', 'Freddie', 'Robert' and 'Peter' and anonymous.
21 (for example, Stahl, 2010; Enloe, 1983; Detraz, 2012; Cohn, 2013).
22 Interview with 'Peter' and 'David'.
23 (Hansard, 2015b).
24 (Hicks Stiehm, 1982; Young, 2003).
25 (Lee, 2018; Philipps, 2012; Maurer, 2015; Donnelly, 2005; Lindlaw, 2008).
26 Interview with 'Tom'.
27 The specific effects on the families (wives, partners, children) is outside the remit of this study, but it forms the basis of my post-doctoral project.
28 Interview with 'Tom' and 'Geoff'.
29 Interview with 'Robert'.
30 This subtitle is taken from the title of the newspaper article of the same name (Abe, 2012).
31 (Gatens, 1999; Grosz, 1994; Pettman, 1997).
32 Interview with 'Geoff'.
33 Interview with 'Steve'.
34 Interview with 'Geoff' and 'Robert'.
35 (Ouma *et al.*, 2011; Chappelle *et al.*, 2014; Thompson, 2006).
36 Interview with 'Robert'.
37 Interview with 'David'.
38 Interview with 'Anonymous'.
39 Interview with 'Tom', 'Amy', 'Steve' and 'David'.
40 Interview with 'Geoff'.
41 (Robinson, 2009; French and McCain, 2005; Talbot, 2012).
42 (Hooper, 1998; Kunashakaran, 2016).
43 (Kunashakaran, 2016; Holmqvist, 2013).

References

Abe, N. (2012) Dreams in infrared: the woes of an American drone operator. *Spiegel Online*, 14 December. Available at: www.spiegel.de/international/world/pain-continues-after-war-for-American-drone-pilot-a-872726.html.

Allsep, L.M. (2013) The myth of the warrior: martial masculinity and the end of don't ask, don't tell. *Journal of Homosexuality*, 60 (2–3): 381–400. doi: 10.1080/00918369.2013.744928.

Anzaldúa, G. (2012) *Borderlands: la frontera: the new Mestiza*. 4th edn. San Francisco, CA: Aunt Lute Books.

Black, B. (2013) *A rare insight into a day at work for a pilot of a drone*. Available at: www.bbc.co.uk/news/world-24522150 (accessed: 27 January 2015).

Blackhurst, R. (2012) The air force men who fly drones in Afghanistan by remote control. *Telegraph* [online], 24 September. Available at: www.telegraph.co.uk/news/uknews/

defence/9552547/The-air-force-men-who-fly-drones-in-Afghanistan-by-remote-control.html (accessed: 19 August 2016).

Braudy, L. (2005) *From chivalry to terrorism: war and the changing nature of masculinity*. 1. Vintage Books edn. New York: Vintage Books.

British Army (2014) Be the best. Available at: www.youtube.com/watch?v=sOhvjBwuix4 (accessed: 01 November 2016).

Buse, P. and Stott, A. (1999) *Ghosts: deconstruction, psychoanalysis, history*. New York: St Martin's Press.

Chappelle, W.L., McDonald, K.D., Prince, L., Goodman, T., Ray-Sannerud, B. and Thompson, W. (2014) Symptoms of psychological distress and post-traumatic stress disorder in United States Air Force 'drone' operators. *Military Medicine*, 179 (8S): 63–70. doi: 10.7205/MILMED-D-13-00501.

Chow, D. (2013) Drone wars: pilots reveal debilitating stress beyond virtual battlefield. *LiveScience*, 5 November. Available at: www.livescience.com/40959-military-drone-war-psychology.html.

Coddington, K.S. (2011) Spectral geographies: haunting and everyday state practices in colonial and present-day Alaska. *Social & Cultural Geography*, 12 (7): 743–756. doi: 10.1080/14649365.2011.609411.

Cohn, C. (ed.) (2013) *Women and wars*. Cambridge, UK; Malden, MA: Polity Press.

Cole, C. (2017) Interview of former RAF Reaper pilot 'Justin Thompson' (a pseudonym) by Chris Cole. Drone Wars UK, May 2017. Available at: https://dronewarsuk.files.wordpress.com/2017/05/justin-thompson-interview-transcript.pdf (accessed: 5 May 2018).

Cole, C. (2018) Interview of Air Marshal Greg Bagwell by Chris Cole. Drone Wars UK. Available at: https://dronewars.net/interview-of-air-marshall-greg-bagwell-drone-wars-uk/ (accessed: 12 November 2018).

Daggett, C. (2015) Drone disorientations: how 'unmanned' weapons queer the experience of killing in war. *International Feminist Journal of Politics*, 17 (3): 361–379. doi: 10.1080/14616742.2015.1075317.

Dawson, G. (1994) *Soldier heroes: British adventure, empire, and the imagining of masculinities*. London; New York: Routledge.

Derrida, J. (2006) *Specters of Marx: the state of the debt, the work of mourning and the New International*. Routledge classics. 1. publ. New York: Routledge.

Detraz, N. (2012) *International security and gender*. Dimensions of security. Cambridge, UK; Malden, MA: Polity.

Donnelly, S. (2005) Long-distance warriors. *TIME*, pp. 42–43.

Ehrenreich, B. (1998) *Blood rites*. St Ives, Great Britain: Virago.

Elshtain, J.B. (1995) *Women and war*. University of Chicago Press edn. Chicago, IL: University of Chicago Press.

Enloe, C.H. (1983) *Does khaki become you?: the militarisation of women's lives*. London: Pluto Press.

Fertout, M., Jones, N. and Greenberg, N. (2012) Third location decompression for individual augmentees after military deployment. *Military Medicine*, 62 (3): 188–195.

Freeman, C. (2015) Armchair killers: life as a drone pilot. *Telegraph*, 4 November. Available at: www.telegraph.co.uk/culture/film/film-news/11525499/good-kill-drone-pilot-true-story.html (accessed: 30 June 2015).

French, S.E. and McCain, J. (2005) *The code of the warrior: exploring warrior values past and present*. 1. paperback edn. Lanham, Md.: Rowman & Littlefield.

Garber, B. and Zamorski, M. (2012) Evaluation of a third-location decompression program for Canadian forces members returning from Afghanistan. *Military Medicine*, 177 (4): 397–403.

Gatens, M. (1999) Power, bodies, and difference. In *Feminist theory and the body: a reader*. Edinburgh: Edinburgh University Press.

Good kill (2015) Film. Niccol, A. 4 October 2015.

Gordon, A. (2008) *Ghostly matters: haunting and the sociological imagination*. New University of Minnesota Press edn. Minneapolis, MN: University of Minnesota Press.

Gothie, S.C. (2015) Stag head matriarchy: postfeminist domesticity and antlers in Domino Magazine 2005–9. *Home Cultures*, 12 (1): 5–28. doi: 10.2752/175174215X141719151 60254.

Green, G., Emslie, C., O'Neill, D., Hunt, K. and Walker, S. (2010) Exploring the ambiguities of masculinity in accounts of emotional distress in the military among young ex-servicemen. *Social Science and Medicine*, 71: 1480–1488.

Grosz, E.A. (1994) *Volatile bodies: toward a corporeal feminism*. Theories of representation and difference. Bloomington: Indiana University Press.

Hansard (2015a) *Armed drones*. House of Commons. Available at: www.publications. parliament.uk/pa/cm201516/cmhansrd/cm151201/halltext/151201h0001.html.

Hansard (2015b) *Drones in conflict*. House of Commons. Available at: www.publications. parliament.uk/pa/cm201516/cmhansrd/cm151013/debtext/151013-0004.htm#151013-0004.htm_spnew20.

Hicks Stiehm, J. (1982) The protected, the protector, the defender. *Women's Studies International Forum*, 5 (3/4): 367–376.

Holmqvist, C. (2013) Undoing war: war ontologies and the materiality of drone warfare. *Millennium – Journal of International Studies*, 41 (3): 535–552. doi: 10.1177/0305829813483350.

Hooper, C. (1998) Multiple masculinities in international relations. In Zalewski, Marysia and Parpart, Jane (eds) *The 'man question' in international relations*. Boulder, CO: Westview Press.

Horn, D.M. (2010) 'Boots and bedsheets: constructing the military support system in a time of war.' In Sjoberg, L., Via, S., and Enloe, C. (eds) *Gender, war and militarism: feminist perspectives*. Santa Barbara, CA: Praeger.

House of Commons, (2008) *Medical care for the armed forces: seventh report of session 2007–2008*. HC 327. Available at: www.publications.parliament.uk/pa/cm200708/cmselect/cmdfence/327/327.pdf (accessed: 10 March 2016).

Hughes, J., Earnshaw, M., Greenberg, N., Eldridge, R., Fear, N., French, C., Deahl, M. and Wessely, S. (2008) Use of psychological decompression in military operational environments, *Military Medicine*, 173 (6): 534–538.

Kaurin, P.M. (2014) *The warrior, military ethics and contemporary warfare: Achilles goes asymmetrical*. Military and defence ethics. Farnham, Surrey: Ashgate Publishing Company.

Kenway, J. (2006) *Haunting the knowledge economy*. London; New York: Routledge.

Killeen, D. (2015) *BBC Five Live Adrian Chiles interviews UK drone pilots*. Available at: www.bbc.co.uk/programmes/b060xx5w#auto (accessed: 7 September 2015).

Kunashakaran, S. (2016) Un(wo)manned aerial vehicles: an assessment of how unmanned aerial vehicles influence masculinity in the conflict arena. *Contemporary Security Policy*, 37 (1): 31–61. doi: 10.1080/13523260.2016.1154405.

Lee, P. (2018) *Reaper Force: the inside story of Britain's drone wars*. London: John Blake Publishing Ltd.

Lindlaw, S. (2008) Remote-control warriors suffer battle stress at a distance Psychologists, chaplains ease mental strain. *Boston News*, 8 August. Available at: www.boston.

com/news/nation/articles/2008/08/08/remote_control_warriors_suffer_battle_stress_
at_a_distance/ (accessed: 22 March 2015).

MacKenzie, J.M. (1997) *The empire of nature: hunting, conservation, and British imperialism*. Studies in imperialism. Manchester, UK; New York: Distributed exclusively in the USA by St Martin's Press.

Mangan, J.A. (2015) *Militarism, hunting, imperialism: blooding the martial male*. Place of publication not identified: Routledge.

Manjikian, M. (2014) Becoming unmanned: the gendering of lethal autonomous warfare technology. *International Feminist Journal of Politics*, 16 (1): 48–65.

Martin, M.J. (2010) *Predator: the remote-control air war over Iraq and Afghanistan: a pilot's story*. Minneapolis, MN: Zenith Press.

Masters, C. (2010) Cyborg soldiers and militarised masculinities. *Eurozine* [online], 20 May. Available at: www.eurozine.com/articles/2010-05-20-masters-en.html (accessed: 3 July 2016).

Maurer, K. (2015) She kills people from 7,850 miles away. *The Daily Beast*, 18 October. Available at: www.thedailybeast.com/articles/2015/10/18/she-kills-people-from-7-850-miles-away.html (accessed: 2 February 2016).

Mayer, J. (2009) The Predator war: what are the risks of the CIA's covert drone program? *The New Yorker*, 26 October. Available at: www.newyorker.com/magazine/2009/10/26/the-predator-war (accessed: 28 April 2019).

Mead, C. (2014) A rare look inside the Air Force's drone training classroom. *The Atlantic*, 4 June. Available at: www.theatlantic.com/technology/archive/2014/06/a-rare-look-inside-the-air-forces-drone-training-classroom/372094/ (accessed: 4 August 2015).

Mulrine, A. (2012) Drone pilots: why war is also hard for remote soldiers. *Christian Science Monitor* [online], 28 February. Available at: www.csmonitor.com/USA/Military/2012/0228/Drone-pilots-Why-war-is-also-hard-for-remote-soldiers (accessed: 1 April 2013).

Niva, S. (1998) Tough and tender: new world order masculinity and the Gulf War. In *The 'man question' in international relations*. Boulder, CO: Westview Press. pp. 109–128.

Ouma, J.A., Chappelle, W.L., and Salinas, A. (2011) *Facets of occupational burnout among US Air Force active duty and National Guard/Reserve MQ-1 Predator and MQ-9 Reaper operators*. 88ABW-2011–4485. Air Force Research Laboratory: Aerospace Medicine Education.

Peterson, Nolan (2016) 'Deep targets': on the ground with British, US drone forces attacking ISIS. *The Daily Signal*. Available at: http://dailysignal.com/2016/05/02/deep-targets-on-the-ground-with-british-u-s-drone-forces-targeting-isis/ (accessed: 19 August 2016).

Peterson, V.S. (2014) Family matters: how queering the intimate queers the international. *International Studies Review*, 16 (4): 604–608. doi: 10.1111/misr.12185.

Pettman, J.J. (1997) Body politics: international sex tourism. *Third World Quarterly*, 18 (1): 93–108. doi: 10.1080/01436599715073.

Philipps, D. (2012) *Lethal warriors: when the new band of brothers came home*. First Palgrave Macmillan paperback edition. New York: Palgrave Macmillan.

Richardson, A. (2012) 'Riding like Alexander, hunting like Diana': gendered aspects of the medieval hunt and its landscape settings in England and France. *Gender & History*, 24 (2): 253–270. doi: 10.1111/j.1468-0424.2012.01681.x.

Robinson, P. (2009) *Military honour and the conduct of war*. Routledge Military Studies. London: Routledge.

Royakkers, L. and van Est, R. (2010) The cubicle warrior: the marionette of digitalized warfare. *Ethics and Information Technology*, 12 (3): 289–296. doi: 10.1007/s10676-010-9240-8.

Skaar, O. (2014) Killing from afar: the growing debate over rewards for drone pilots. *Curiousmatic*. Available at: http://curiousmatic.com/killing-afar-growing-debate-rewards-drone-pilots/ (accessed: 4 January 2015).

Stahl, R. (2010) *Militainment, Inc.: war, media, and popular culture*. New York: Routledge.

Talbot, S. (2012) *Warriors, warfighting and the construction of masculine identities*. In Brisbane, 2012. Available at: www.tasa.org.au/wp-content/uploads/2012/11/Talbot-Steven.pdf (accessed: 19 May 2015).

The Economist (2014) Dilbert at war: the stressful lives of the 'Chair Force'. *The Economist*, 21 June.

Thompson, W. (2006) *Effects of shift work and sustained operations: operator performance in remotely piloted aircraft*. HSW-PE-BR-TR-2006–0001. Air Force Research Laboratory.

Walker, S. (2010) Assessing the mental health consequences of military combat in Iraq and Afghanistan: a literature review: soldiers mental health, *Journal of Psychiatric and Mental Health Nursing*, 17 (9): 790–796.

Young, I.M. (2003) The logic of masculinist protection: reflections on the current security state. *Signs*, 29 (1): 1–25.

Zamorski, M., Guest, K., Bailey, S. and Garber, B. (2012) Beyond battlemind: evaluation of a new mental health training program for Canadian forces personnel participating in third-location decompression, *Military Medicine*, 177 (11): 1245–1253.

Conclusion

'Life is [still] complicated'[1]

That 'life is complicated' is something that social science sometimes neglects to take seriously as a theoretical statement (Gordon, 2008, p. 3). The complexity of social life is so obvious as to appear almost not worth noting but is actually extremely important. Through the lives and experiences of Reaper crews, I have illuminated the need for deploying the framework of Haunting as a meaningful way of engaging with this complexity. The scholarship on military technologies, such as drones, has tended to focus on singular reasons for the importance (or not) of these developments rather than acknowledging the multiplicity of factors at work. For example, there has been a marked tendency to produce caricatures and to provide limited data that fails to represent the crews as the complex, inter-esting and varied individuals they are. In reading those articles I was struck by the way the crews were made mono-dimensional, and by the way that the social, psychological, historical and cultural context in which they operated was absent. Therefore, through my interviews with ex-Reaper crews I reflected on just how much more these individuals had to say than was currently available in the aca-demic literature and popular press, and am delighted to see more thorough and detailed data finally becoming available (see for example, Lee, 2018).

One of the reasons that Singer argues that drones are 'the most important weapons development since the atomic bomb' is because they change the way that people interact on the battlefield (Singer, 2011, p. 10). It is not *just* that the crews have the capacity to kill from a distance at very little reciprocal risk (that has been done before), nor is it *just* that they can conduct persistent surveillance operations for hugely extended periods of time (this has largely been the purview of intelligence agents), nor is it *just* that they cycle to and from the war space and the home space with such bizarre regularity. It is rather the interweaving of all three. And this has important implications for how gender intersects with the phenomenon of drone warfare – as outlined throughout the book.

Perhaps one of the reasons that airpower specialists often seem somewhat baffled by the media frenzy that has accompanied the inclusion of armed drones into the Royal Air Force is that the aeroplane itself is not *that* exciting, nor *that* novel. Sure, some of the technology that can be strapped onto the Reaper is interesting, but the physical body of the aeroplane (what is known as the air frame) is pretty standard. What this perspective misses, and I think is one of the reasons that drones have so clearly captured the public imagination, is that the interest is not in the technology itself. But rather in the experiences of the human beings using that technology. In a world that has seen James Bond(s) (and similar characters) wield all manner of awesome, bizarre, and complex technologies in publicly consumable ways, a remote-control aeroplane is rather low grade. But the experiences of the person who *conducts war* by remote control aeroplane makes for a good story.

As such, I had two primary reasons for choosing British Reaper crews as my case study. First, as an empirical case, it is greatly understudied and dotted with detail and intricacies. And second, following on from this, within the discourses on armed drone use there was a lack of understanding of the differences, similarities, nuances and peculiarities of this case in comparison with that of the other major user of armed drones like Reaper, the United States. It has felt to me during this process that the complexity of understanding warfare, and of understanding how innovations in technology effect warfare, is part and parcel of understanding the vast array of reasons that people have for engaging in warfare or working for militaries. The complexity of the war technologies/gender intersection is mirrored by the variety of experiences that these crews (who represent a small subsection of one arm of the British armed forces) and the different ways that they reflected back on their time within either 13 or 39 squadron. It is also reflected in the way that they *felt* about engaging with an airframe that represents both 'cutting edge' technology and a source of social derision.

By using the framework of Haunting in this book I have aimed to engage with all of this complexity. Far from trying to discipline the data to fit into categories or to fit a particular narrative, Haunting has allowed me to interact with the pieces that do not fit: the outliers and the contradictions. As a poststructural feminist scholar, I was most interested in the gendered arguments about drones, the way that, whether arguing for the feminisation or masculinisation of the crews, previous accounts always seemed to include elements more suited to the opposing perspective. The arguments about masculinisation acknowledged the increasing need for skills previously considered feminine, and arguments about feminisation noted elements, such as emotional distance, which fit in with narratives about masculinisation. Through these pieces, just as much as the pieces that settle and sediment in the ways that we expect them to, it is possible to engage with the core ideas of Haunting – the importance of the absent (as much as the present), the invisible (as much as the visible), the dead (as much as the alive). Haunting situates individuals within their specific context, acknowledging the important and multifaceted way in which history/ies and myths haunt us in ways that we can sometimes understand and sometimes cannot; and the ways in which the future looms over today woven through with hopes, dreams and fears. One of the ways in which Haunting enables this kind of interweaving is through an engagement with 'sensuous knowledges'. These knowledge(s), as outlined in Chapter 2, are the kinds of things that exceed the traditional kinds of data used in social science; for example, they are the hunches, the symbols, the intuitions, the memories, and the body language. By using these kinds of knowledge(s), I have sketched out a portrait of British Reaper crews that illuminates some of the ways in which their lives contradict themselves, fold in on themselves, make sense read forwards and read in reverse. Therefore, this book represents less a sequential collection of chapters and perhaps more a collage, an almost unwieldy mind map, linked by the red threads of gender. Whilst I *began* with just interview data from British Reaper crews, it became clear that there was also more to say about how this data agreed, disagreed, and was the different/same/similar to other accounts – journalists' accounts, memoirs, films, cartoons, and parliamentary debates. And so, I have included these throughout the book.

Fitting the collage into an organised structure meant at times drawing what felt like arbitrary lines in spaces where these did not exist, but rather blurred and blended into one another. There are references in Chapter 5 that would have fitted equally well in Chapter 6, and simultaneously there are quotes in Chapter 4 that contradict other quotes that I've placed in Chapter 4. And yet, through all of this I had the sense of something, the shape of something that even in its translucence seemed to me to make sense. What I have tried to write, then, is something that draws attention to the fact that coherence is only partial, and relies, at least in part, on sensuous knowledge.

Reflecting on the (cob)webs of gender led me to make an amendment to the framework of Haunting through the addition of Cynthia Weber's queer logic(s). Haunting includes a concern with a range of different binaries: absence and presence, silence and scream, life and death. Additionally, spectralities scholars point to the liminal spaces in between these two categories. However, I felt an absence within Haunting itself. I have added queer logic(s) to Haunting as a vocabulary that is able to explicitly draw attention to the binary destabilising capabilities of the ghosts. As I outlined in Chapter 2, queer theory is concerned with the disciplinary power of dichotomous thinking. To feminism's concerns with male/female and masculine/feminine queer theorists add the binaries of normal/perverse and heterosexual/homosexual (amongst others).[2] One of the ways in which queer theory adds to thinking about, and challenging, binaries is through the adoption of logics which deny the need to choose *between* the two poles, arguing instead for the possibility of multiple identities, destabilising the 'slash' at the heart of the dyads.[3]

I add queer logic(s) to the methodology of the ghost hunt because, as I argued in Chapter 2, Haunting is already troubled by entities that embody this logic. By making this addition, I render explicit the possibilities encompassed in the ghostly (alive *at the same time* as dead, present *at the same time* as being absent), always already 'exceeding all binary opposites' (Harris, 2015, p. 17). Highlighting the possibilities of thinking beyond binaries makes it possible to speak about military technologies *simultaneously* masculinising *and* feminising the crews; makes it possible to speak about the ways in which the crews are haunted by the histories of the past *and* the future *at the same time*: 'actual present: now', the 'spectral moment' (Derrida, 2006, p. xix).

Ghostly warriors

Utilising the ghost hunt, I have investigated the gendered narratives that emerge through the various iterations of 'the warrior'. Initially it felt somewhat anachronistic to include references to *The Iliad* and the Knights Templar in a project centred on examining the effects of cutting edge military technology. The Reaper and the technology it can be laden with is a far, far cry from the slingshots used by the ancients or the plate armour and lances that knights staggered under. And yet, within contemporary narratives of war this/these figure(s) flickered and seethed around the edges. References to military 'brothers' and the need to protect 'innocents' continue to (re)inscribe what it *means* to be part of the British Armed Forces. I have tried to illustrate that the historical/mythical legacy of what it means to conduct war, and what makes an individual valuable in warfare, has implications for how users of novel military technologies, like drones, are understood. The ghost(s) of the warrior reveals himself to be not the history of one man, but rather a collection of translucent layers of histories, myths, odes, and art. Engaging in conversation with this ghost, I argue that the power of the masculine as hierarchically empowered over the feminine has important repercussions for the individuals who conduct drone warfare today. Similarly, the relationship between the warrior and the means (technologies) of war are also illuminated as woven through with gendered stories about physical strength, cunning, valour and courage which continue to pervade the debate. For example, arguments about cowardice (outlined in Chapter 4) are informed by the bravery required to engage in hand-to-hand combat whether with swords, spears, or physical wrestling. Similarly, the caricatures of drone crews (see Chapter 5) as overweight and childish are woven through with myths that construct warriors as physically strong and emotional stoical *men*.

The warrior also reaffirms the importance of engaging with sensuous knowledges: the emotional, the sensuous, the subconscious. After all, the derision of Reaper drone crews does not spring from a concern that the crews cannot do their jobs, or do their jobs well, but rather from a *sensation*, that is hard to articulate, that the jobs that they do are somehow lacking. That the very act of crewing an armed drone, whilst operationally useful, is somehow *less* than crewing a manned aircraft. The ghostliness of the warrior figure similarly emerges in the idea that military masculinity is constructed around the experience of *being there*, physically in the battlefield (outlined in Chapter 6). Therefore, whilst the aircraft that crews fly is present above the battlefield, the human beings, their bodies, are not. This disjuncture between being and not being serves as a reminder of the way in which the crews embody the queer logic that I have used in conjunction with Haunting. The ghostly figure of the warrior has haunted, and discomforted both the crews and those who comment on their actions, revealing the instability and irrationality of military masculinity. I say 'irrationality' because the masculinity that the crews are being either denied or (re)inscribed with often relies on the same 'data' as 'evidence', and the assessment of which pole the crews may situate themselves (masculinised or feminised) depends on knowledges that are beyond the empirical. Therefore, I have tried to embrace these strange knowledges and the inherent contradictions in the world of the Reaper crews because, rather than trying to force them to fit into the masculine *or* feminine box, I have used Weber's queer logic to argue for the capacity for the apparently opposing 'sides' to co-exist – or rather that these experiences destabilise the categories entirely.

Whilst I presented the warrior in Chapter 3 through the 'traditional' lens of heteronormative masculinity, I hope that through the inclusion of this figure in the subsequent chapters it became clear that this ghostly shape is far more complicated than the initial reading of the stories suggested. I produced this 'straight' reading because I felt it most accurately reflected the way(s) that the warrior figure is currently utilised and understood in contemporary discourses about the military and within the military. However, as with the construction of masculinity itself, what constitutes valuable behaviour in the military has changed, remodelled, and moved in line with new technologies – be that gun powder, intercontinental-ballistic-missiles, or cyber viruses.[4] But again, as with masculinity/ies, the *concept* of the warrior remains useful, coherent, and surprisingly powerful.

For the subsequent chapters I engaged in a ghost hunt of the lives of British Reaper crews, using the marsh lights from the trail of the warrior to see their experiences through the prism of Haunting. Dwelling in the curious liminal spaces the ghost provides, I conjured the sensuous knowledges associated with the core concerns of Haunting: (1) complex personhood, (2) in/(hyper)visibility, (3) disturbed temporality, and (4) power.

These four components of Haunting are woven through my analysis of the lives of British Reaper crews and through the complexity that surrounds the entire debate on the use of armed drones. The way in which the Reaper crews can kill has implications for how they are perceived by their colleagues and also by outsiders – press and public, for example. The symbolic importance of killing to constructions of masculinity is confused by the lack of risk that the crews face as they undertake lethal missions. This apparent risklessness (in common narratives, but which I illustrated in Chapter 4 is only partially true) serves to feminise the crews as 'protected' where the capacity to kill emphasises their masculinity. The resulting confusion of gendered discourses is one of the reasons that there is so much discomfort surrounding what Reaper crews do. Their lives refuse to fit into our neat categories of masculine or feminine. But rather, they reflect a queer logic(s) that exceeds both of these categories, much as the warrior exceeds the label of 'warrior', and the experience of war exceeds 'war' itself.[5] Drawing attention to the humans engaged in drone warfare draws attention to the complexity of personhood, to the limitations of the existing masculine/feminine categories, and our discomfort with the blurring of boundaries in interesting ways that are not possible in discussions that focus solely on the legal, strategic, or mechanical issues.

The second component of Haunting, that of in/(hyper)visibility is drawn out through an exploration of the persistent surveillance role of the Reaper crews.

The inherent spectrality in looking and being looked at maps onto the gendered binary that situates the masculine as the subject and observer, and the feminine as the object(ified) and watched. Challenging the ideas of masculinist ocularcentricity that dates from the Enlightenment, I utilised sensuous knowledges as an alternative that is better able to capture nuance, detail and the things that cannot easily be seen, but which help to paint a more complete picture of the ways in which gender is working in the lives of British Reaper crews. Referring to interviewees' hunches, hopes, and fears as well as references from popular culture such as films and cartoons, I have argued for the importance of the inclusion of subjective data. This has included sensations of 'wrongness', indistinct shapes, and movements in the corners of the eye. Such data enabled me to explore how the crews have been rendered both invisible and *hyper*visible in narratives about the use of drones and to engage with their perspectives on this in/(hyper)visibility. The ghost hunt disturbed the separation between the categories of masculine 'watcher' and feminine 'watched'. It did so by illuminating the ways in which the experience of watching can be equally as feminising as it is masculinising. Appealing to queer logic(s), the experience of being watched can provide a platform through which to demonstrate military masculinity whilst *at the same time* eroding (and thereby feminising) the warrior identity constructed through professional trust.

Disturbed temporality, the third component of Haunting, emerges clearly through the Reaper crews' experience of flying night shifts that are *simultaneously* day shifts, and being absent from a theatre of war that they are also present in. The history and myths of the warrior from the past mingle with hopes and fears of the future to infuse the discourses of the present. What it means to be a warrior *today* (in what Derrida refers to as the 'actual present: now' as related to the past present and the future present (2006, p. xix)) is influenced by these different threads and shades. Similarly, what it meant to be a warrior (historically) *then* and what it might mean to be a warrior (future) *then* is implicated in what it means to be a warrior *now*. Through the prism of Haunting I have argued that the Reaper crews' experiences of cycling between the war space (coded masculine) and the home space (coded feminine) is dangerously disturbing as it results in the bleeding of the barriers between the two, illustrating the way this dislocation haunts the crews as they try to make sense of who they are when and where, then and there.

Power, the final component of Haunting, draws together the three previous components and illuminates the way that this framework can usefully be applied to feminist concerns about the way(s) that gender hierarchies dis/empower. Each of the above components engages with questions of whether individuals and groups are dis/empowered. For example, how does the complexity of personhood serve to discipline individuals and reduce their agency? And to what extent does acknowledging their complexity provide more nuanced means of understanding which can empower? In Chapter 5, I cited feminist connections between visibility and emancipation – the political necessity of being seen in order to be heard. How then, does the way that the Reaper crews have been first, written out of the debate about drones and then written in as dangerous, deviant, exotic 'Others' restrict or otherwise affect the way the individuals involved perceive themselves and their roles? How is the history and mythology behind the figure of the warrior haunting the contemporary narratives that both celebrate and critique the use of military technologies such as armed drones? How can we better understand the way that concern about the loss of valour can exist in the same cultures that are risk averse in warfare? By using Haunting, it is possible to raise all these questions as spectres – to conjure them into conversation and explore how the gendered dichotomies at work in narratives about military technologies create around the (cob)webs of power – that restrict, discipline and increase agency in interesting and unexpected ways.

The framework of Haunting is integral to the exploration of the gendered dynamics of security studies. Through this framework, I have engaged with the complexity at the heart of the lives of British Reaper crews, providing new insights into the debate about military technology and gender. I have augmented the methodology of the ghost hunt with 'queer logic(s)' to further illuminate the way in which the narratives about these crews *simultaneously* destabilise and (re)inscribe dominant understandings of military masculinity. The addition of queer logic(s) makes it possible to draw the concerns of Haunting (the complexity of personhood, in/(hyper)visibility, disturbed temporality and power) into closer alignment with the thinking of feminist scholarship. As such, this innovative combination provides the nuance and detail that is vital to the investigation of the entanglement of military technologies and gender.

Notes

1 (Williams, 1995, p. 10).
2 (Sedgwick, 2008; Ahmed, 2006).
3 (Weber, 2014, 2016).
4 (Coker, 2007; Singer, 2011; Bousquet, 2008).
5 (After Scarry, 1987).

References

Ahmed, S. (2006) *Queer phenomenology*. Durham, NC: Duke University Press.
Baggiarini, B. (2015) Drone warfare and the limits of sacrifice. *Journal of International Political Theory*, 11 (1): 128–144. doi: 10.1177/1755088214555597.
Coker, C. (2007) *The warrior ethos: military culture and the war on terror*. LSE international studies series. London; New York: Routledge.
Derrida, J. (2006) *Specters of Marx: the state of the debt, the work of mourning and the New International*. Routledge classics. 1. publ. New York: Routledge.
Gordon, A. (2008) *Ghostly matters: haunting and the sociological imagination*. New University of Minnesota Press edn. Minneapolis, MN: University of Minnesota Press.
Harris, V. (2015) Hauntology, archivy and banditry: an engagement with Derrida and Zapiro. *Critical Arts*, 29 (sup1): 13–27. doi: 10.1080/02560046.2015.1102239.
Lee, Peter (2018) *Reaper force: the inside story of Britain's drone wars*. London: John Blake Publishers.

Scarry, Elaine (1987) *The body in pain: the making and unmaking of the world*. New York: Oxford University Press.

Sedgwick, E. (2008) *Epistemology of the closet*. Berkeley, CA: University of California Press.

Singer, P.W. (2011) *Wired for war: the robotics revolution and conflict in the 21st century*. London: Penguin.

Weber, C. (2014) From queer to queer IR. *International Studies Review*, 16 (4): 596–601. doi: 10.1111/misr.12160.

Weber, C. (2016) *Queer international relations: sovereignty, sexuality and the will to knowledge*. Oxford studies in gender and international relations. New York: Oxford University Press.

Williams, P. (1995) The alchemy of race and rights. Cambridge, MA: Harvard University Press.

Index